REWARD

REWARD

Elementary

Teacher's Book

Simon Greenall

MACMILLAN HEINEMANN
English Language Teaching

Macmillan Education
Between Towns Road Oxford OX4 3PP, UK
A division of Macmillan Publishers Limited

Companies and representatives throughout the world.

ISBN 0 435 24242 3

Designed by Stafford & Stafford
Cover design by Stafford & Stafford
Cover illustration by Martin Sanders

Illustrations by:
Adrian Barclay (Beehive Illustration), pp27, 78
Paul Beebee (Beehive Illustration), pp36, 37
Gillian Hunt (Beehive Illustration), pp4, 38, 92, 100, 101
Sarah McDonald, pp94, 103
Martin Sanders, pp11, 13, 20, 25, 27, 29, 32, 34/35, 37, 46, 47, 49, 58, 60, 64, 73, 76, 80, 84, 87, 88, 90, 91, 97, 98, 99, 101
Simon Smith, pp31, 59, 61, 64, 85, 99
Stafford & Stafford pp22, 23

Commissioned photography by:
Chris Honeywell pp10/11, 14, 40/41, 44, 45, 50, 54, 62, 94, 102
p5 ©HELT; p16 ©HELT(x3); M.Van Gelderen p100

Author's Acknowledgements
I am very grateful to all the people who have contributed towards *Reward* Elementary. Thank you so much to:
- All the teachers I have worked with on seminars around the world, and the various people who have influenced my work.
- James Richardson for the happy and efficient work he has done on producing the tapes, and the actors for their voices.
- Philip Kerr, for his comments on the material, which are especially helpful and well-considered.
- Sue Side, Maria Zeny, Sue Bailey, Sue Watson, Marcella Banchetti,Taska Eszter, Steve Bilsborough, Etienne Andre, Brian Waine, Burt Johnson, Miguel Almarza and Jim Scrivener for their reports on the material. I have tried to respond to all their suggestions, and if I have not always been successful, then the fault is minealone.
- Simon Stafford for his usual, skilful design.
- Douglas Williamson for his efficient design management.
- Pippa McNee for tracking down some wonderful photos.
- Catherine Smith for her support and advice and her sensitive management of the project.
- Angela Reckitt for her attention to detail, her contribution to the effectiveness of the course, and her calm, relaxed style which makes work such a pleasure.
- and last, but by no means least, Jill, Jack and Alex.

Acknowledgements
The author and publisher would like to thank the following for their kind permission to reproduce material in this book: Heinemann Educational, a division of Reed Educational and Professsional Publishing Limited, for an extract from *Dear Ruth...Love Jan* by N. McIver; Reed Books Limited, for an extract from *The Return of Heroic Failures* by Steven Pile, published by Secker & Warburg Limited, 1988; Times Publishing Group, Singapore, for an extract from *Culture Shock! USA* by Esther Wanning, published by Times Editions.

Photographs by: Anthony Blake Photo Library/Phototeque Culinaire pp80, Anthony Blake p105, Art Directors & Trip Photographic Library pp3, 20, 74, 103, British Rail International p57; Carlton TV p26; Corbis p86; European Passenger Services photograph p28; Eye Ubiquitous p74; Ronald Grant Collection p43; Chris Honeywell p33; Hulton Getty p50; Image Bank pp16, 30, 39, 70, 71, 76, 82, 88, 103; Images Colour Library pp7, 16, Lake School of English pp8, 9; Louis Psihoyos/Material World p18; Jeremy Hartly/Panos Pictures p15; Pictor International-London pp2, 103, 105, Rex Features p39, South American Pictures/Tony Morrison p2; Frank Spooner Collection p42; Stockphotos p30; Tony Stone Images pp2, 16, 19, 20, 30, 47, 56, 69, 70, 74, 82, 88, 89, 103, 105; Superstock p105; Universal Pictorial Press p39; Zefa Pictures p30

The publishers should also like to thank Celia Bingham, Nick Blinco, British Rail International, Carlton TV, Tim Cater, Aleeta Cliff, Tim Friers, Louis Harrison, Sue Kay at The Lake School of English, Helen Kidd, Sarah McDonald, Jamie McNee, Jason Mann, Andrew Oliver, Philip and Valerie Opher, Anthony Reckitt, Rebecca Smith, Martha and Rebecca Stafford, Julie Stone, Lydia Trapnell, Douglas Williamson, Chris Winter and Verley Woodley.

While every effort has been made to trace the owners of copyright material in this book, there have been some cases when the publishers have been unable to contact the owners. We should be grateful to hear from anyone who recognises their copyright material and who is unacknowledged. We shall be pleased to make the necessary amendments in future editions of the book.

Printed and bound in Spain by Mateu Cromo, S.A. Pinto (Madrid)

2003 2002 2001

13 12 11 10 9 8

Contents

Introduction

Course organisation

Reward is a general English course which will take learners of English from their first introduction to English up to a fully proficient use of the language. British English is used as the model for grammar, vocabulary, spelling and pronunciation, but other varieties of English are included for listening and reading practice. The course components for each level are as follows:

For the student	For the teacher
Student's Book	Teacher's Book
Practice Book	Class cassettes
Practice Book cassette	Resource Pack
	Business Resource Pack

The Student's Book has forty teaching lessons and Progress check lessons. After every eight teaching lessons there is a Progress check lesson to review the language covered in the preceding teaching lessons and to present new language work relevant to the grammar, functions and topics covered so far. Within the teaching lessons the main grammar or language functions and the most useful vocabulary are presented in boxes which allow easy access to the principal language of the lesson. This makes the focus of the lesson clearly accessible for purposes of presentation and revision. Each lesson will take between 90 and 120 minutes.

The Class cassettes contain all the listening and sounds work in the Student's Book.

The Practice Book has forty practice lessons corresponding to the forty teaching lessons in the Student's book. The activities are designed for self-access work and can be easily used to provide extra practice for the language presented in the Student's Book, either in the class or as self-study material. Each lesson will take between 60 and 80 minutes to do.

The Practice Book cassettes contain all the listening and sounds work in the Practice Book.

The Resource Packs provide extra teaching material to practise the main language points of the teaching lessons. There is a general Resource Pack for learners of general English and a Business Resource Pack for learners with language requirements of a more professional nature. The material is photocopiable. *Reward* is designed to be very flexible in order to meet the very different requirements of learners and the course. So the Resource Packs can be used to extend a core teaching lesson of 90 – 120 minutes from the Student's Book with approximately 45 minutes of extra material for use in the classroom.

The Teacher's Book contains a presentation of the course design, methodological principles, detailed teaching notes interleaved with pages from the Student's Book and four photocopiable tests. The teaching notes for each lesson include a step-by-step guide to teaching the lesson, a discussion of some of the difficulties the learners may encounter, and more detailed methodological issues arising from the material presented.

Course design

The course design is based on a broad and integrated multi-syllabus approach. It is broad in the sense that it covers grammar and language functions, vocabulary, reading, listening, speaking, writing, and sounds explicitly, and topics, learner training and socio-cultural competence implicitly. It is integrated in that each strand of the course design forms the overall theme of each lesson. The lessons always include activities focusing on grammar and language functions, and vocabulary. They will also include reading, listening, speaking, writing and sounds. The inclusion of each strand of the syllabus is justified by its communicative purpose within the activity sequence. The methodological principles and approaches to each strand of course design are discussed below.

Methodological principles

Here is an outline of the methodological principles for each strand of the course design.

Grammar and language functions

Many teachers and learners feel safe with grammar and language functions. Some learners may claim that they want or need grammar, although at the same time suggesting that they don't enjoy it. Some teachers feel that their learners' knowledge of grammar is demonstrable proof of language acquisition. But this is only partly true. Mistakes of grammar are more easily tolerated than mistakes of vocabulary, as far as comprehension is concerned, and may be more acceptable than mistakes of socio-cultural competence, as far as behaviour and effective communication is concerned. *Reward* attempts to establish grammar and language functions in its pivotal position but without neglecting the other strands of the multi-syllabus design.

Vocabulary

There are two important criteria for the inclusion of words in the vocabulary boxes. Firstly, they are words which the elementary learner should acquire in order to communicate successfully in a number of social or transactional situations. The revised Threshold Level (1990)[1] has been used to arbitrate on the final selection. Secondly, they may also be words which, although not in Threshold Level, are generated by the reading or listening material and are considered suitable for the elementary level. However, an overriding principle operates: there is usually an activity which allows learners to focus on and, one hopes, acquire the words which are personally relevant to them. This involves a process of personal selection or grouping of words according to personal categories. It is hard to acquire words which one doesn't need, so this approach responds to the learner's individual requirements and personal motivation. *Reward* Elementary presents approximately 950 words in the vocabulary boxes for the learner's active attention, but each learner must decide which words to focus on.

Reading

The reading passages are generally at a higher level than one might expect for learners at elementary level. Foreign language users who are not of near-native speaker competence are constantly confronted with difficult language and to expose the learners to examples of real-life English in the reassuring context of the classroom is to help prepare them for the conditions of real life. There is always an activity or two which encourages the learner to respond to the passage on a personal level or to focus on its main ideas. *Reward* attempts to avoid a purely pedagogical approach and encourages the learner to respond to the passages in an authentic way before using it for other purposes.

Listening

Listening is based on a similar approach to reading in *Reward.* Learners are often exposed to examples of natural, authentic English in order to prepare them for real-life situations in which they will have to listen to ungraded English. But the tasks are always graded for the learners' particular level. Furthermore, a number of different native and non-native accents are used in the listening passages, to reflect the fact that in real life, very few speakers using English speak with British or American standard pronunciation.

Speaking

Many opportunities are given for speaking, particularly in pairs and groupwork. Learners are encouraged to work in pairs and groups, because the number of learners in most classes does not allow the teacher to give undivided attention to each learner's English. In these circumstances, it is important for the teacher to evaluate whether fluency or accuracy is the most important criterion. On most occasions in *Reward* Elementary speaking practice in the *Grammar* sections is concerned with accuracy, and in the *Speaking* sections with fluency. In the latter case, it is better not to interrupt and correct the learners until after the activity is ended.

Writing

The writing activities in *Reward* are based on guided paragraph writing with work on making notes, turning notes into sentences, and joining sentences into paragraphs with various linking devices. The activities are quite tightly controlled; this is not to suggest that more creative work is not valid, but it is one of the responsibilities of a coursebook to provide a systematic grounding in the skill. More creative writing is covered in the Practice Book. Work is also done on punctuation, and most of the writing activities are based on real-life tasks, such as writing letters, newspaper articles, a report of a debate etc.

Sounds

Pronunciation, stress and intonation work tends to interrupt the communicative flow of a lesson, and there is a temptation to leave it out in the interests of maintaining the momentum of an activity sequence. In *Reward* there is work on sounds in most lessons, usually just before the stage where the learners have to use the new structures orally in pair or group work. At this level, it seems suitable to introduce work beyond the straightforward system of English phonemes, most of which the learners will be able to reproduce accurately because the same phonemes exist in their own language. So activities which focus on stress in words and sentences, and on the implied meaning of certain intonation patterns are included. The model for pronunciation is standard British.

Topics

The main topics proposed by the Threshold Level 1 are covered in *Reward* Elementary. These include personal identification, house and home, daily life, leisure activities, travel, relations with other people, health, education, shopping, food and drink, geographical location and the environment. On many occasions the words presented in the vocabulary box all belong to particular word fields or topics.

Learner training

Implicit in the overall approach is the development of learner training to encourage learners to take responsibility for their own learning. Examples of this are regular opportunities to use mono- and bi-lingual dictionaries, ways of organising vocabulary according to personal categories and inductive grammar work.

Cross-cultural training

Much of the material and activities in *Reward* create the opportunity for cross-cultural training. Most learners will be using English as a medium of communication with be using English as a medium of communication with other non-native speakers, and certainly with people of different cultures. Errors of socio-cultural competence are likely to be less easily tolerated than errors of grammar or lexical insufficiency. But it is impossible to give the learners enough specific information about a culture, because it is impossible to predict all the cultural circumstances in which they will use their newly acquired language competence. Information about *sample* cultures, such as Britain, America as well as non-native English speaking ones, is given to allow the learners to compare their own culture with another. This creates opportunities for learners to reflect on their own culture in order to become more aware of the possibility of different attitudes, behaviour, customs, traditions and beliefs in other cultures. In this spirit, cross-cultural training is possible even with groups where the learners all come from the same cultural background. There are interesting and revealing differences between people from the same region or town, or even between friends and members of the same family. Exploring these will help the learners become not merely proficient at the language but competent in the overall aim of communication.

Level and progress

One important principle behind *Reward* is that the learners arrive at elementary level with very different language abilities and requirements. Some may find the early units very easy and will be able to move quickly onto later lessons. These learners can confirm that they have acquired a certain area of grammar, language function and vocabulary, consolidate this competence with activities giving practice in the other aspects of the course design, and then move on. Others may find that their previous language competence needs to be reactivated more carefully and slowly. The core teaching lesson in the Student's Book may not provide them with enough practice material to ensure that the given grammar, language functions and vocabulary have been firmly acquired. For these learners, extra practice may be needed and is provided in both the Practice Book (for self-study work) and the Resource Packs (for classroom work). If learners return to language training at elementary level after a long period of little or no practice, it is hard to predict quite what they still know. *Reward* is designed to help this kind of learner as much as those who need to confirm that they have already acquired a basic knowledge of English.

Interest and motivation

Another important principle in the course design has been the intrinsic interest of the materials. Interesting material motivates the learners, and motivated learners acquire the language more effectively. The topics have been carefully selected so that they are interesting to adults and young adults, with a focus on areas which would engage their general leisure-time interests. This is designed to generate what might be described as authentic motivation, the kind of motivation we have when we read a newspaper or watch a television programme. But it is obvious that we cannot motivate all learners all of the time. They may arrive at a potentially motivating lesson with little desire to learn on this particular occasion, perhaps for reasons that have nothing to do with the teacher, the course or the material. It is therefore necessary to introduce tasks which attract what might be described as pedagogic or artificial motivation, tasks which would not usually be performed in real life, but which engage the learner in an artificial but no less effective way.

Variety of material and language

Despite the enormous amount of research done on language acquisition, no one has come up with a definitive description of how we acquire either our native language or a foreign language which takes account of every language learner or the teaching style of every teacher. Every learner has different interests and different requirements, and every teacher has a different style and approach to what they teach. *Reward* attempts to adopt an approach which appeals to differing styles of learning and teaching. The pivotal role of grammar and vocabulary is reflected in the material but not at the expense of the development of the skills or pronunciation. An integrated multi-syllabus course design, designed to respond to the broad variety of learners' requirements and of teachers' objectives, is at the heart of *Reward*'s approach.

The Threshold Level, 1990, J. A van Ek and JLM Trim, Council of Europe Press, 1991

RESEARCH

Macmillan Education is committed to continuing research into coursebook development. Many teachers contributed to the evolution of *Reward* through piloting and reports, and we now want to continue this process of feedback by inviting users of *Reward* - both teachers and students - to tell us about their experience of working with the course. If you or your colleagues have any comments, queries or suggestions, please address them to the Publisher, Adult Group, Macmillan Education, Between Towns Road, Oxford, OX4 3PP or contact your local Macmillan representative.

Map of the book

Lesson	Grammar and functions	Vocabulary	Skills and sounds
1 *What's your name?* Introducing yourself	Present simple (1): *to be* Asking and saying who people are and where they are from	Countries	**Reading:** reading for main ideas **Listening:** listening for main ideas **Sounds:** syllable stress **Speaking:** asking and saying who you are and where you are from
2 *This is Bruno and Maria* Jobs and nationalities	Present simple (2): *to be* Possessive adjectives Articles (1): *a/an*	Jobs	**Sounds:** syllable stress **Reading:** reading for main ideas **Writing:** writing a description saying who people are, where they come from and what their job is
3 *Questions, questions* Personal information	Questions Negatives Short answers	Forms giving personal information	**Listening:** listening for specific information **Speaking:** using a model dialogue for guided speaking practice **Writing:** completing a form; joining two pieces of information with *and*
4 *How many students are there?* A brochure for a language school	*There is/are* Plurals (1): regular Position of adjectives	Numbers Classroom vocabulary Adjectives to describe people or things	**Sounds:** pronunciation of numbers **Reading:** reading for specific information **Speaking:** talking about your ideal school **Writing:** writing a brochure for your ideal school
5 *Where's my pen?* Personal possessions	*Has/have got* Prepositions of place (1)	Personal possessions	**Listening:** listening for context; listening for specific information **Sounds:** pronunciation of the alphabet **Speaking:** talking about personal possessions
Progress check lessons 1–5	Revision	Organising new vocabulary Expressions to use when you don't understand	**Sounds:** /ɪ/ and /iː/; syllable stress **Reading:** reading for main ideas **Writing:** punctuation of sentences
6 *Families* Family relationships	Possessive *'s* and *s'* Plurals (2): regular and irregular	Members of the family	**Sounds:** weak syllables /ə/ **Reading:** predicting; reading for main ideas; inferring **Speaking:** talking about your family
7 *What's the time?* Customs around the world at different times of day	Present simple (3): for customs and routines Prepositions of time (1)	Times of day Meals	**Speaking:** telling the time **Reading:** reading for main ideas **Listening:** listening for specific information **Writing:** writing an informal letter
8 *Home* The home of a family in Russia	*Some* and *any* (1) Prepositions of place (2)	Rooms Furniture Parts of the house	**Sounds:** syllable stress **Reading:** reading for specific information **Listening:** predicting; listening for specific information **Writing:** completing a chart about different homes
9 *How do you relax?* Different ways of relaxing	Present simple (4): for habits and routines *Wh-* questions Third person singular (*he/she/it*)	Hobbies	**Reading:** reading for specific information **Sounds:** third person singular present simple endings, /s/, /z/,/ɪz/ **Listening:** listening for specific information **Speaking:** talking about ways of relaxing in your free time
10 *Do you like jazz?* Different kinds of entertainment	Pronouns Present simple (5): talking about likes and dislikes	Entertainment and recreation	**Reading:** reading for main ideas; reading for specific information **Listening:** listening for main ideas; listening for specific information **Sounds:** interested intonation in sentences **Writing:** writing an advert for a penfriend
Progress check lessons 6–10	Revision	Parts of speech Noting down new vocabulary Features of a dictionary	**Sounds:** /ə/; /s/ and /z/; polite intonation **Reading:** reading for main ideas **Speaking:** talking about lifestyles in Britain and in your country

Lesson	Grammar and functions	Vocabulary	Skills and sounds
11 *A day in my life* A day in the life of a TV presenter	Present simple (6): saying how often you do things Prepositions of time (2)	Everyday actions	**Reading:** reading for main ideas; understanding text organisation **Speaking:** talking about your typical day **Listening:** listening for specific information **Writing:** writing about a typical day in your life using *then, after that* and *and*
12 *How do you get to work?* Different means of transport	Articles (2): *a/an, the* and no article Talking about travel	Means of transport Adjectives and their opposites	**Reading:** reading for main ideas; reading for specific information **Sounds:** /ðə/ and /ðiː/ **Speaking:** talking about how to get to work/school **Writing:** writing a paragraph on how to get to work/school
13 *Can you swim?* Talking about what people can and can't do	*Can* and *can't* Questions and short answers	Skills and abilities	**Reading:** reading and answering a questionnaire **Sounds:** strong and weak forms of *can* **Listening:** listening for specific information; listening for context **Speaking:** talking about personal abilities
14 *How do I get to Queen Street?* Finding your way around a town	Prepositions of place (3) Asking for and giving directions	Shops and town facilities	**Listening:** listening for specific information; listening for main ideas **Reading:** predicting; reading for main ideas; reading for specific information **Speaking:** talking about shopping in Britain and your country
15 *What's happening?* Describing actions taking place at the moment	Present continuous	Compass points Adjectives to describe places	**Listening:** listening for main ideas; listening for specific information **Sounds:** stressed words **Speaking:** talking about cities around the world **Writing:** writing a postcard to a friend
Progress check lessons 11–15	Revision	International words Words from other languages Words and expressions used in everyday situations	**Sounds:** /ð/ and /ə/; /ɑː/; /æ/ and /ʌ/; syllable stress; stressed words **Writing:** writing addresses **Speaking:** asking for and giving addresses
16 *Who was your first friend?* Talking about childhood	Past simple (1): *be*	Adjectives to describe character	**Speaking:** talking about childhood **Listening:** listening for main ideas; listening for specific information **Sounds:** strong and weak forms of *was* **Writing:** writing true and false statements about your childhood
17 *How about some oranges?* Typical food at different meals	*Some* and *any* (2) Countable and uncountable nouns Making suggestions	Items of food and drink	**Listening:** listening for specific information; listening for main ideas **Sounds:** /s/ and /z/ in plural nouns; /ən/ in noun expressions with *and* **Speaking:** talking about eating conventions in Britain and your country
18 *I was born in England* Biographies of Sting and Whitney Houston	Past simple (2): regular verbs *Have*	Words connected with the music business Important life events	**Reading:** predicting; reading for specific information **Sounds:** past simple endings: /t/, /d/ and /ɪd/ **Listening:** predicting; listening for specific information **Speaking:** talking about the life of a famous person
19 *What's he like?* Talking about appearance	Describing people (1)	Adjectives to describe people's appearance Physical features	**Writing:** writing a letter giving a personal description **Listening:** listening for main ideas; listening for specific information **Speaking:** talking about typical appearance

Lesson	Grammar and functions	Vocabulary	Skills and sounds
***20** A grand tour* Visiting cities in Europe	Past simple (3): irregular verbs *Yes/no* questions and short answers	Verbs connected with tourism Irregular verbs	**Sounds:** pronunciation of some European cities and countries **Listening:** listening for main ideas; listening for specific information **Speaking:** talking about a holiday
Progress check lessons 16–20	Revision	Word association Word charts Words formed from other parts of speech	**Sounds:** /ɔː/, /ɜ/ and /ɒ/; silent letter patterns; interested intonation in sentences; syllable stress **Speaking:** playing a game called Twenty Questions
***21** Mystery* An extract from an article about Agatha Christie	Past simple (4): negatives *Wh-* questions	New words from a passage about Agatha Christie	**Reading:** predicting; reading for main ideas; reading for specific information **Sounds:** weak stress of auxiliary verbs **Writing:** writing a short autobiography **Speaking:** talking about your life
***22** Dates* Special occasions and important dates	Past simple (5) Expressions of time	Ordinal numbers Dates, months of the year	**Sounds:** pronunciation of ordinal numbers and dates **Listening:** listening for main ideas; understanding text organisation **Speaking:** talking about your answers to a quiz
***23** What's she wearing?* Clothes and fashion	Describing people (2) Present continuous or present simple	Items of clothing and accessories Colours Actions of the face, hand and body	**Listening:** listening for specific information **Reading:** reading and answering a questionnaire **Speaking:** discussing your answers to the questionnaire
***24** I'm going to save money* Talking about resolutions and future plans	*Going to* *Because* and *so*	New verbs from this lesson	**Reading:** predicting; reading for main ideas **Listening:** predicting; listening for main ideas **Writing:** writing resolutions; joining sentences with *because* and *so*
***25** Eating out* Eating in different kinds of restaurants	*Would like* Talking about prices	Food items	**Listening:** predicting; listening for specific information **Sounds:** pronunciation of *like* and *'d like* **Reading:** reading for specific information **Writing:** writing a paragraph about eating out in your country
Progess check lessons 21–25	Revision	Techniques for dealing with difficult words Word puzzle	**Sounds:** words with the same pronunciation but different spelling; /ʊ/ and /uː/; syllable stress; polite intonation **Writing:** preparing a quiz **Speaking:** asking and answering quiz questions
***26** Can I help you?* Shopping	Reflexive pronouns Saying what you want to buy Giving opinions Making decisions	Items of shopping	**Listening:** listening for specific information; listening for main ideas **Sounds:** polite and friendly intonation **Speaking:** talking about shopping habits
***27** Whose bag is this?* Describing objects and giving information	*Whose* Possessive pronouns Describing objects	Adjectives to describe objects Materials	**Speaking:** describing objects; acting out conversations in a Lost Property office **Listening:** listening for main ideas; listening for specific information
***28** What's the matter?* Minor illnesses Healthcare in Britain	Asking and saying how you feel Sympathising *Should, shouldn't*	Adjectives to describe how you feel Nouns for illnesses Parts of the body	**Listening:** listening for main ideas **Reading:** reading for specific information; dealing with unfamiliar words **Writing:** writing an advice leaflet about healthcare for visitors in your country
***29** Country factfile* Facts about Thailand, United Kingdom and Sweden	Making comparisons (1): comparative and superlative forms of short adjectives	Adjectives to describe countries Measurements	**Sounds:** syllable stress; pronunciation of measurements **Reading:** reading for specific information **Listening:** listening for specific information **Writing:** writing a factfile for your country

Lesson	Grammar and functions	Vocabulary	Skills and sounds
30 *Olympic spirit* Olympic records	Making comparisons (2): comparative and superlative forms of longer adjectives	Sports Adjectives to describe sports	**Listening:** listening for main ideas **Speaking:** talking about Olympic sports **Reading:** understanding text organisation
Progress check lessons 26–30	Revision	Adjectives and opposites Word Zigzag	**Sounds:** words which rhyme; /ei/ and /aɪ/; stress patterns **Speaking:** conversation building for sentences **Listening:** listening for main ideas
31 *When in Rome, do as the Romans do* Customs and rules in different countries	*Needn't, can, must, mustn't*	New words from this lesson	**Reading:** reading for main ideas **Speaking:** talking about rules and customs in different countries **Listening:** listening for main ideas; listening for specific information **Sounds:** pronunciation of *must* and *mustn't* **Writing:** writing advice and rules for visitors to your country
32 *Have you ever been to London?* Travel experiences	Present perfect (1): talking about experiences	New words from this lesson	**Reading:** reading for main ideas **Sounds:** strong and weak forms of *have* and *haven't* **Speaking:** talking about things to do in your town; talking about experiences **Writing:** writing a postcard
33 *What's happened?* Talking about good and bad luck	Present perfect (2): talking about recent events *Just* and *yet*	Adverbs and their opposites	**Listening:** listening for specific information **Speaking:** talking about things you have and haven't done; acting out conversations about good and bad luck
34 *Planning a perfect day* Favourite outings	Imperatives Infinitive of purpose	Words connected with outings	**Speaking:** talking about a perfect day **Reading:** reading for main ideas **Writing:** writing advice for planning the perfect day out
35 *She sings well* Schooldays	Adverbs	Adverbs and their opposites	**Sounds:** identifying attitude and mood **Reading:** reading for main ideas; reading for specific information **Listening:** listening for main ideas **Speaking:** talking about achievement at school; performing an action in the style of different adverbs
Progress check lessons 31–35	Revision	Collocation	**Sounds:** words with the same vowel sound; /əʊ/ and /ɔɪ/; word stress and a change of meaning **Reading:** reading for specific information **Writing:** focusing on unnecessary words
36 *I'll go by train* Travel by train and plane	Future simple (1): (*will*) for decisions	Features of an airport and station	**Listening:** listening for specific information **Reading:** reading for specific information **Speaking:** acting out a role-play in a travel agent's
37 *What will it be like in the future?* Talking about the future	Future simple (2): (*will*) for predictions	Nouns and adjectives for the weather	**Listening:** listening for specific information **Reading:** predicting; reading for specific information
38 *Hamlet was written by Shakespeare* World facts	Active and passive	Verbs used for passive	**Speaking:** talking about true and false sentences **Reading:** reading and answering a quiz **Listening:** listening for specific information **Writing:** writing a quiz about your country
39 *She said it wasn't far* Staying in a youth hostel	Reported speech: statements	Items connected with travel	**Reading:** reading for main ideas **Listening:** listening for specific information **Writing:** writing a letter of complaint
40 *Dear Jan ... Love Ruth* A short story by Nick McIver	Tense review	New words from this lesson	**Reading:** predicting; reading for main ideas **Listening:** listening for specific information **Writing:** writing a different ending to the story
Progress check lessons 36–40	Revision	Prepositions Word association	**Sounds:** /ɔ/ and /ɔɪ/; syllable stress **Speaking:** playing *Reward Snakes and Ladders*

1 What's your name?

Present simple (1): *to be*; asking and saying who people are and where they're from

READING AND LISTENING

1 Match the sentences and the people in the photos.

a Hello, I'm Lah. I'm from Thailand.
b Hello, I'm Renato. I'm from Brazil.
c Hello, I'm Sally. I'm from the United States of America.
d Hello, I'm Mehmet. I'm from Istanbul.

2 Read and listen.

M Hello, I'm Marco. What's your name?
E Hello, Marco. My name's Enrique.
M Where are you from, Enrique?
E I'm from Madrid.

3 Put the sentences in the right order to make a conversation.

☐ I'm George. Where are you from, Marie?
☐ I'm from Paris. And you?
☐ I'm from Athens.
☐ Hello. What's your name?
☐ Hello, my name's Marie. What's your name?

4 Work in pairs and check your answers to 3.

Now listen and check.

1

GENERAL COMMENTS

Reward Elementary
This level of the *Reward* series assumes that the learners have already had some formal instruction in English, or have been exposed to the language at some time in the past. The students will be encouraged to reactivate their passive knowledge of the language, but they will also be given the opportunity to study again the basic elements of English grammar and vocabulary. Point out to them that the target grammar and vocabulary of each lesson is clearly signposted in the two boxes: in this lesson, the main focus is on the present simple (1): *to be*, asking and saying who people are and where they come from, and on the towns and countries in the *Vocabulary and sounds* section.

Names

This first lesson naturally focuses on names, and before it begins, you need to be clear what you would like to call your students and what you would like them to call you. Should it be family names or first names? The answer is whichever is the most socio-culturally appropriate.

READING AND LISTENING

1 Aim: to prepare for listening; to present the target structures.

• Point to yourself and say who you are and where you're from. If you have any photos of famous people, hold them up in front of you and pretend to be these people. For the moment, it isn't necessary to ask the students to use the structures for themselves. Until they have more confidence, they can just listen and absorb the sounds of English.

• Ask the students to match the sentences and the people. They should look for physical clues to perform this activity.

Answers
a photo - bottom middle
b photo - left
c photo - top middle
d photo - right

2 Aim: to practise reading and listening.

• [cassette] This exchange is the target conversation for this lesson. Ask students to read and listen as you play the tape. You may play the tape several times.

• Use the exchange for pronunciation practice if you think the students are ready to speak. Role play the conversation with one or two students, then ask them to work in pairs and continue to act it out.

3 Aim: to practise reading for text organisation.

• This kind of scrambled sentence activity allows the students to focus on the internal logic of a piece of spoken or written English, why one sentence goes in one position but rarely in another. Ask the students to read the sentences and number them in the order in which they go.

• Don't give the answer until they have listened to the conversation in activity 4.

• Ask pairs of students to act out the conversation to the whole class.

4 Aim: to practise listening; to check comprehension.

• [cassette] Ask the students to listen and check their answers to activity 3.

Answers
1 Hello. What's your name?
2 Hello, my name's Marie. What's your name?
3 I'm George. Where are you from, Marie?
4 I'm from Paris. And you?
5 I'm from Athens.

GRAMMAR AND FUNCTIONS

1 Aim: to present the present simple of the verb *to be*; to focus on word order.

• Ask the students to read the information in the grammar box and then to do the activities.

• The students will already have seen the target structure in the *Reading and listening* activities. This activity is designed to focus on its components and their position. Ask the students to write the sentences out with the words in the right order. You can also do this activity orally in pairs.

Answers
1 Hello, I'm Manuel.
2 Where are you from?
3 What's your name?
4 I'm from London.
5 Hello, my name's Toni.
6 My name's Hannah.

● You may need to check your students' punctuation. You may also need to explain that the apostrophe *'s* and *'m* is the abbreviated form of *is* and *am*, and both are not only perfectly acceptable but extremely common in spoken and informal written English.

2 Aim: to practise using the target structure.
● Explain that the sentences in activity 1 are two sides of an exchange. Ask the students to match the sentences.

Answers
1 – 5, 2 – 4, 3 – 6

● You may like to explore other possible matches.

3 Aim: to practise asking and saying who people are and where they're from.
● Ask the students to do this activity orally in pairs.

Answers
1 What is your name?
2 Where are you from?
3 I'm from China.
4 My name is Peter.
5 Hello, I'm Johann.
6 Where are you from?

VOCABULARY AND SOUNDS

1 Aim: to present the English words for some common towns and countries.
● Write the words *cities* and *countries* on the board and write a few words in the correct column. Ask students to suggest where the other words should go. Continue with other cities and countries which are not in the box. Try to focus on those places, such as neighbouring cities, towns and countries, which you're likely to use often in class.

2 Aim: to focus on syllable stress.
● You may like to tell your students that English is a stress-timed language, in which the stress can change from syllable to syllable, and not a syllable-timed language, where there is equal stress on each syllable. Tell them if your language is stress- or syllable-timed.

● Ask the students to say the words aloud as they listen to the tape. Ask the students to count the number of syllables in each word.

Answers
two syllables: Britain, Bangkok, Japan
three syllables: Budapest,Tokyo, Canada

3 Aim: to focus on syllable stress.
● Ask the students to say the words aloud and underline the stressed syllable.

● Play the tape and ask the students to check their answers.

Answers
London Mexico Thailand Hungary Australia Brazil

4 Aim: to focus on the spelling and pronunciation of the students' own towns and countries.
● If the words for the students' towns and countries are very different in pronunciation and spelling, take some time in getting the students to say them aloud.

5 Aim: to practise the pronunciation of the target structures.
● Play the tape again and ask the students to listen and repeat the target structures.

SPEAKING

1 Aim: to practise using the target structures.
● Ask the students to ask and say who they are and where they're from. They should do this with as many people as possible. Encourage them to get up and walk around the classroom, if this is appropriate or possible.

2 Aim: to practise using the target structures.
● Make a list of famous people on the board, and ask the students to suggest additions to it.

● Ask the students to say where the famous people are from, and write these places by the names.

● Ask the students to 'become' one of these famous people, and to continue practising the structures learnt in this lesson, by going round and saying who they are and where they're from.

3 Aim: to check comprehension.
● Ask the students to write down the names of the famous people they have met.

● If this is the first time the students have worked together, it might be a good idea for them to write a seating plan of the class with the names of every student. If they can't remember everyone's name, they can go and ask him or her again.

GRAMMAR AND FUNCTIONS

> Present simple (1): *to be*
> ***I'm*** *Lah. (= I am Lah.)*
> ***My name's*** *Enrique. (= My name is Enrique.)*
> ***What's*** *your name? (= What is your name?)*
>
> Asking and saying who people are
> *What's your name?* *My name's Anna.*
> *I'm Anna.*
>
> Asking and saying where people are from
> *Where are you from?* *I'm from Italy.*

1 Put the words in the right order and write sentences.

1 I'm hello Manuel.
2 you are from where?
3 your name what's?
4 London from I'm.
5 name's hello my Toni.
6 Hannah name's my.

2 Match the sentences in 1.

Hello, I'm Manuel. *Hello, my name's Toni.*

3 Complete the sentences.

1 What _____ your name?
2 Where are you _____?
3 I _____ from China.
4 My name _____ Peter.
5 Hello, _____'m Johann.
6 _____ are you from?

1 What is your name?

VOCABULARY AND SOUNDS

1 Look at the words in the vocabulary box. Put them in two columns: *cities* and *countries.*

> Britain Bangkok Tokyo Budapest Japan Canada
> London Mexico Thailand Hungary Australia Brazil

cities: Bangkok *countries: Britain*

2 Listen and repeat these words.

□□ Britain
□□□ Tokyo Canada
□□ Bangkok
□□□ Budapest
□□ Japan

How many syllables does each word have?

Britain, two *Tokyo, three ...*

3 Look at these words and underline the stressed syllable.

London Mexico Thailand Hungary
Australia Brazil

Listen and check.

4 What's your country's name in English?

5 Listen and repeat.

name your name What's your name?
name My name's Pat.
from where Where are you from?
from I'm from Britain.

SPEAKING

1 Ask and say who you are and where you're from. Use the conversation in *Reading and listening* activity 2 and 3 to help you.

2 Think of a famous person and where they come from. Go round asking and saying who you are and where you come from.

Hello, I'm Bill Clinton. What's your name?
Hello, Bill. I'm Nelson Mandela. Where are you from?
I'm from the United States of America.

3 Write the names of the people you meet and where they come from.

2 This is Bruno and Maria

Present simple (2): ***to be;*** **possessive adjectives; articles (1):** ***a/an***

He's a journalist.

He's an _____.

She's a police officer.

She's a _____.

She's an artist.

He's a _____.

VOCABULARY AND SOUNDS

1 Complete the sentences with words from the box.

artist police officer receptionist secretary teacher waiter doctor engineer student farmer journalist

2 Turn to Communication activity 10 on page 100 and check you know what the other jobs are.

3 Listen and look at the stressed syllables.

artist journalist secretary receptionist engineer

4 Listen and repeat these words.

teacher waiter doctor farmer student

5 Work in pairs. Underline the stressed syllables in the words in 4.

Listen and check.

READING

1 Look at the photos. Where do you think the people are from, *Britain, Brazil* or *Japan*?

2 Match the photos and the descriptions below them.

3 Answer the questions.

1 Where is Pete from?
2 What's his job?
3 Where are Bruno and Maria from?
4 What's their job?
5 Where is Michiko from?
6 What's her job?

2

GENERAL COMMENTS

Course design

Remind the students that *Reward* has two principal syllabuses: grammar/functions and vocabulary, and a series of secondary syllabuses: reading, writing, speaking, listening and sounds. Every lesson will contain the principal syllabuses. There are usually ten or twelve words in each vocabulary box. These are the words which are considered suitable to be learnt at this level, and at this stage of the course. There may be other words which occur, and which the students want to note down, but you should limit this, as too much vocabulary will simply be forgotten. The secondary syllabuses are often combined with one another, which reflects the fact that it is rare for them to be used in complete isolation.

Jobs

Many words for jobs are similar in two languages, so make a list of these before you come to the class.

VOCABULARY AND SOUNDS

1 Aim: to present some of the words in the vocabulary box.

- The words in the box are some of the most common jobs, but you may wish to supplement this list with jobs which are personally relevant to the students, such as their own or their parents' occupations.

- Ask the students to complete the sentences with words from the box.

Answers
He's an engineer.
She's a doctor.
He's a waiter.

- You may like to reinforce the process of acquiring these words by asking the students to categorise them under different headings, for example, high/low status, high/low pay, male/female etc.

2 Aim: to present the rest of the words in the vocabulary box.

- Ask the students to turn to the Communication activity and check they know what the other jobs are.

3 Aim: to focus on syllable stress in words.

- Ask the students to say the words aloud. For the moment, don't correct their pronunciation.

- Play the tape. Ask the students to listen and check their pronunciation and the placing of the stress. This will force them to listen more attentively to the pronunciation.

4 Aim: to focus on the pronunciation of the words.

- Play the tape again. Ask the students to listen and repeat the words.

- When you've played the tape several times, ask individual students to read out the list of words to the rest of the class.

5 Aim: to focus on stressed syllables and the /ə/.

- Ask the students to work in pairs and to underline the stressed syllables in the words in 4.

- Write the words on the board and ask individual students to mark the stress on the words.

- Play the tape and ask the students to check their answers.

- Tell the students that the unstressed syllable in many English words is the /ə/ sound. If they stress the syllable which the /ə/ sound represents, their foreign pronunciation will be immediately noticed.

Answers
teacher waiter doctor farmer student

READING

1 Aim: to prepare for reading.

- Explain that it is often useful to prepare for reading in a foreign language by making predictions about what you're about to read. It's something we do naturally and quickly in our language but which may need to be done more consciously in a foreign language.

- Ask the students to predict where the three people might be from.

Answers
Brazil - 1
Japan - 2
Britain - 3

2 Aim: to practise reading for main ideas.

- This micro skill is an important one to develop, particularly when the students deal with more complex passages. This activity is very simple, but will give practice in a kind of reading which will become more and more common as the learning process proceeds.

Answers
A - 3 B - 1 C - 2

3 Aim: to check comprehension.

- These questions are designed to encourage the students to take a closer look at the passages.

- You can do this activity orally with the whole class, or as pairwork.

Answers
1 Pete's from London.
2 He's an engineer.
3 Bruno and Maria are from São Paulo.
4 They're doctors.
5 Michiko is from Tokyo.
6 She's a teacher.

- Point out that in reply to the question *Where are you from?* the answer can be either a town or a country. Both take the preposition *from*.

GRAMMAR

1 Aim: to focus on the forms of the verb *to be*.

- Ask the students to read the information in the grammar box and then to do the exercises.

- This activity encourages the students to think about the form of the verb *to be*. It acknowledges the fact that students sometimes mix up *am, is* and *are*, but focuses on the limits of the potential confusion.

Answers
Three: *am, is, are*

2 Aim: to focus on questions and answers.

- You can check this activity orally with the whole class.

Answers
1 d 2 a 3 e 4 c 5 b

3 Aim: to focus on possessive adjectives and contracted forms of *to be*.

- Students may confuse possessive adjectives and the contracted forms of *to be*, so this activity highlights the potential confusion.

Answers
1-b 2-b 3-b 4-a

4 Aim: to focus on the use of articles with jobs; to focus on the use of *a* and *an*.

- Tell the students that the indefinite article is always required before a job. Ask students to think of a job and go round asking and saying what their jobs are.
I'm a businessman.
NOT *I'm businessman.*

- Make sure the students use *an* before words which begin with a vowel.

Answers
1 an 2 a 3 an 4 a 5 a 6 a

WRITING

1 Aim: to practise writing; to practise using the target structures.

- This is the first writing activity of *Reward* Elementary and you may want to tell the students about your marking policy. Decide if you're going to encourage accuracy or fluency in writing, or both at different times. Show them the marks you use to indicate wrong words, words in the wrong position, missing words, spelling mistakes, grammar mistakes etc.

- Ask the students to write descriptions of the two people.

Answers
1 This is Petra Schmidt. She's from Hamburg in Germany. She's a student.
2 This is Rosario Barbisan. He's from Montevideo in Uruguay. He's a teacher.

2 Aim: to practise writing; to practise using the target structures.

- Ask the students to write similar paragraphs about people they know.

- You may like to ask the students to do this activity for homework.

A *This is my friend Pete Jenkins. He's from London in Britain. He's an engineer.*

B *This is Bruno and Maria Rodrigues. They're from São Paulo in Brazil. They are doctors.*

C *This is my friend Michiko. She's from Tokyo in Japan. She's a teacher.*

GRAMMAR

Present simple (2): *to be*

I'm *(= I am)*
*you'**re** (= you are)*
*he'**s** (= he is) she'**s** (= she is) it'**s** (= it is)*
*we'**re** (= we are)*
*they'**re** (= they are)*

Possessive adjectives

	my	
What's	*your*	
	his	*name?*
This is	*her*	*friend.*
	our	
	their	

Articles (1): *a/an*

You use *a/an* before jobs.
*I'm **a** student. He's **an** engineer.*

You use *a* before all nouns which begin with a consonant.
What's your job? *I'm **a** student.*

You use *an* before all nouns which begin with a vowel.
What's his job? *He's **an** accountant.*

1 How many different forms are there for the present simple of the verb *to be*?

2 Match the questions and the answers.

1 Where's he from?	a Her name's Fatima.
2 What's her name?	b She's a doctor.
3 Where are they from?	c His name's Andrew.
4 What's his name?	d He's from Brazil.
5 What's her job?	e They're from Argentina.

3 Choose the correct sentence.

1 a His from Brazil.
  b He's from Brazil.
2 a Their from Argentina.
  b They're from Argentina.
3 a He's name is Andrew.
  b His name's Andrew.
4 a Their name's Rodrigues.
  b They're name is Rodrigues.

4 Complete the sentences with *a* or *an*.

1 He's _____ engineer.	4 She's _____ doctor
2 I'm _____ student.	5 He's _____ teacher.
3 She's _____ artist.	6 He's _____ waiter.

WRITING

1 Write a description of the people in the chart. Say who they are, where they're from and what their job is.

Name:	Petra Schmidt	Rosario Barbisan
Town/country:	Hamburg Germany	Montevideo Uruguay
Job:	Student	Teacher

This is Petra Schmidt.
She's from ...

2 Think of someone you know. Write a description. Say who they are, where they're from and what their job is.

Questions, questions

Questions; negatives; short answers

LISTENING AND SPEAKING

1 Read and listen to the first part of this conversation.

JANIE	Hi, Holly, how are you?
HOLLY	Fine thanks, how are you?
JANIE	I'm OK. Who's this?
HOLLY	This is Greg.
JANIE	And how old is he?
HOLLY	He's twenty-three.

2 Put the sentences a – e in the second part of the conversation.

JANIE	What's his job?
HOLLY	(1)_____
JANIE	An actor! What's his surname?
HOLLY	(2)_____
JANIE	Is he from Hollywood?
HOLLY	(3)_____
JANIE	He's very good-looking. What's his phone number?
HOLLY	(4)_____
JANIE	Is he your boyfriend?
HOLLY	(5)_____

a He's an actor. b That's a secret! c Sheppard.
d No, he isn't. He's from New York. e Yes, he is.

Listen and check.

3 Work in pairs. Act out the conversation.

3

GENERAL COMMENTS

Questions

In some languages, it is possible to ask a question by making a statement with a particular intonation. This is possible in English, but it is more common to use a question form with suitable intonation. The word order in questions often causes students difficulty, as does the correct use of the auxiliary. This lesson focuses on questions with the verb *to be*, which can be formed by a question word and/or a simple inversion. Care in establishing the rules and plenty of practice at this stage is advisable if you want your students to avoid common mistakes later on.

Fluency and accuracy

You may like to point out the distinction between fluency and accuracy and how a shift from one to the other will occur at different stages in the lesson and in the course. You may want to point out that, at this early stage, when the students have relatively little language, the language practice will focus mainly on accuracy, but that at a later stage, you will not want to correct them every time they make a mistake, in order to develop their fluency in English.

LISTENING AND SPEAKING

1 Aim: to prepare for listening.

- Explain to the students that they are going to read the first part of a conversation.
- Ask the students to read and listen to the first part and decide who is speaking and if they know each other.
- You may like to ask the students to act out the first part of the conversation.
- Point out that a common response to *How are you?* is *Fine, thanks, how are you*? You usually greet someone with these expressions when you know them or have met them before. For people you meet for the first time, you say *How do you do?* and you reply *How do you do?* Neither form is really a question, more a formulaic exchange.

2 Aim: to prepare for listening; to practise reading for text organisation.

- It would be quite possible simply to give the students the transcript of the conversation they are about to hear, and to follow it as they listen. However, this activity helps the students to focus on the organisation of the conversation, and to prepare for listening by doing some careful reading.
- Ask the students to decide where the sentences go. When they have finished, play the tape and ask them to check their answers.

Answers
1a 2c 3d 4b 5e

3 Aim: to practise speaking.

- Ask the students to act out the conversation in pairs.
- Ask two or three pairs to perform the conversation in front of the whole class.

GRAMMAR

1 Aim: to practise asking questions; to focus on word order in questions.

• Ask the students to read the information in the grammar box and then to do the exercises.

• You may like to do this activity orally with the whole class.

Answers
1 How old are you?
2 Are you a teacher?
3 What's her surname?
4 What's your job?
5 How are your children?
6 What's her address?

• Remind the students that questions always have a question mark at the end of the sentence.

2 Aim: to focus on negatives.

• Ask the students to look back at the conversation in *Listening and speaking* activities 1 and 2, and to mark the statements in this activity true (T) or false (F).

Answers
1T 2F 3T 4F 5T 6T

3 Aim: to practise using questions and short answers.

• Ask the students to match the questions with the answers.

Answers
1b 2a 3d 4c 5e

4 Aim: to practise writing short answers.

• Make sure the students realise that the questions are directed at themselves, and they should answer honestly.

VOCABULARY AND WRITING

1 Aim: to present the words in the vocabulary box.

• The words in the vocabulary box are those which are commonly found on forms, such as Landing Cards, which you often need to complete when you enter a foreign country.

• You may want to write your name and other details on the board, to make the meaning clear. Of course, you don't have to tell them your first name or any other inappropriate information. This may be a suitable moment to teach *Mr, Mrs, Miss* and *Ms* pronounced /mz/.

Answers
1 surname 2 first name 3 job 4 age
5 address 6 phone number 7 married

• You may like to explain that in most English-speaking contexts, if someone asks you your name, you give your first name followed by your surname. Another expression for *surname* in English is *family name.*

• Tell the students that in Britain and America it is very common for people to use first names, even between, for example, a boss and an employee, or an older and a younger person. If they are invited to use first names, they should do so if they feel comfortable with such apparent informality.

2 Aim: to provide a model for writing.

• Ask the students to complete this description with the information on the Landing Card.

Answers
Greg Sheppard is an actor and he's 23. His address is 365 Avenue of the Americas, New York, and his phone number is (212) 693 4428. He isn't married.

• You may like to point out that the brackets around the 212 in the phone number show that you do not need to dial these numbers if you're calling within New York.

3 Aim: to practise writing; to focus on the use of *and*.

• Ask the students to look at the second Landing Card, and at the description in 2. Explain that *and* is used to join together two pieces of information.

• Ask the students to write a description of Fiona North using *and* to join together two pieces of information.

• You may like to ask students to do this activity for homework.

Answers
Fiona North is a teacher and she's 25. Her address is 17 Hill Street, Bristol, and her phone number is (0117) 56731. She is married.

GRAMMAR

> **Questions**
> **You can form a question:**
> **– with a question word.**
> ***What****'s his job?* ***How*** *old is he?*
> **– without a question word.**
> ***Is*** *he from New York?* ***Are*** *you an actor?*
>
> **Negatives**
> ***I'm not*** *(= I am not)*
> *you* ***aren't*** *(= you are not)*
> *he* ***isn't*** *(= he is not) she* ***isn't*** *(= she is not)*
> *it* ***isn't*** *(= it is not)*
> *we* ***aren't*** *(= we are not)*
> *they* ***aren't*** *(they are not)*
>
> **Short answers**
> *Is he your boyfriend? Yes, he is.*
> *Is he from Hollywood? No, he isn't.*
> *Are you an actor? No, I'm not.*
> *Is your surname Sheppard? No, it isn't.*

1 Put the words in the right order and write sentences.

1 old you are how?
2 you a teacher are?
3 her what's surname?
4 what's job your?
5 are children how your?
6 her what's address?

2 Look at the conversation in *Listening and speaking* activities 1 and 2. Are these sentences true or false? Write T for true and F for false.

1 Greg isn't a doctor.
2 He isn't from New York.
3 He isn't twenty-four.
4 He's from Hollywood.
5 His surname is Sheppard.
6 His phone number is a secret.

3 Match the questions and the answers.

1 Is he American?	a No, it isn't.
2 Is his first name Pete?	b Yes, he is.
3 Are you a student?	c Yes, they are.
4 Are they friends?	d Yes, I am.
5 Is she from New York?	e No, she isn't.

4 Write short answers to the questions.

1 Are you from England?	4 Are you sixteen?
2 Are you married?	5 Are you a student?
3 Is your surname Smith?	6 Is your phone number 01278 66554?

1 No, I'm not.

VOCABULARY AND WRITING

1 Complete the missing information on the Landing Card with words from the box.

married	surname	address	age	job	first name	phone number

Landing Card

1 ______	SHEPPARD
2 ______	GREG
3 ______	ACTOR
4 ______	23
5 ______	365 AVENUE OF THE AMERICAS, NEW YORK N.Y 10021
6 ______	(212) 693-4428
7 ______	NO

2 Look at the Landing Card in 1 and complete this description.

____ Sheppard is an ____ and he's ____. His address is 365 Avenue of the Americas, New York, and his ____ is (212) 693-4428. He isn't ____.

3 Look at the Landing Card below.

Landing Card

1 Surname	NORTH
2 First name	FIONA
3 Job	TEACHER
4 Age	25
5 Address	17, HILL STREET, BRISTOL
6 Phone number	0117 56731
7 Married	YES

Now look at the description in 2 again and notice how you join two pieces of information together with *and*.

Greg Sheppard is an actor ***and*** *he's twenty-three.*

Now write a description of Fiona North. Use the description in 2 to help you. Join two pieces of information together with *and*.

How many students are there?

***There is/are*; plurals (1): regular; position of adjectives**

SOUNDS AND VOCABULARY

1 Listen and repeat these numbers.

1 2 3 4 5 6 7 8 9 10 11
12 13 14 15 16 17 18 19 20

2 Write the numbers for these words.

three thirteen thirty four fourteen
forty five fifteen fifty six sixteen
sixty seven seventeen seventy
eight eighteen eighty nine
nineteen ninety one hundred

Now check your answers.
Turn to Communication activity 20 on page 104.

3 Look at the stress pattern of these words.

□	□ □	□ □
three	thirteen	thirty
four	fourteen	forty

Match these words with the stress patterns above.

five fifteen fifty
six sixteen sixty
seven seventeen seventy
eight eighteen eighty
nine nineteen ninety

Now listen and check.

4 Look at the words in the box. Which of the items can you see in the photo?

classroom self-study room café
reception desk computer
cassette player book table chair
car park library

5 Look at the adjectives in the box.

friendly beautiful international
good kind interesting
popular

Which adjectives can you use to describe:
a person, a school, a classroom, a town, a lesson?

a person – friendly, beautiful …

READING

1 Read the brochure for the International English School. Find the answers to these questions.

1 Where's the International English School?
2 Why is it 'international'?
3 How many teachers are there?
4 Where are the students in Kevin's class from?
5 Who is Patricia's teacher?
6 What's the phone number?

2 Work in pairs and check your answers.

Welcome to the International English School in Oxford!

At our school there are two hundred students from thirty countries. There are twelve students in a class, twenty classrooms and thirty teachers. In each modern classroom there are tables, chairs and a cassette player. There are two self-study rooms with computers and a library. There's a café and a car park.

Our students say:

'In my class there are two people from Italy, three from Brazil, five from Japan and one from Greece. It's a very international class!' *Kevin, Malaysia*

'Andrew is a good teacher and he's very popular.' *Patricia, France*

4

GENERAL COMMENTS

Numbers

The presentation and practice of numbers in this lesson assumes that this is not the first time students have encountered them in English. You may want to give a lot more extra practice than there is space for in the Student's Book. Have a look at the practice activities in the Practice Book and the Resource Pack for ideas.

Reading

The passage in this lesson has been carefully written so that all the language is comprehensible to the student at this stage of the course. However, other passages in *Reward* Elementary are at a level slightly higher than your students might expect. This is deliberate and is designed to help them develop their reading skills and their ability to deal with difficult words, such as they will have to do in real-life contexts, in the supportive environment of the classroom.

SOUNDS AND VOCABULARY

1 Aim: to present the pronunciation of numbers 1 – 20.

- Write the numbers 1 - 20 on the board.
- Play the tape and ask the students to listen and repeat the numbers.
- Point to numbers on the board and ask individual students to say what the number is.

2 Aim: to present the written form of numbers.

- Ask the students to write the numbers.
- You may like to ask the students to check their answers in pairs before they turn to the Communication activity.

3 Aim: to focus on the distinction between *-ty* and *-teen*.

- Say the words clearly several times.
- Ask the students to say the words aloud.
- Ask the students to match the stress patterns with the numbers.

Answers
■ five six eight nine
▪■ fifteen sixteen eighteen nineteen
■▪ seven fifty sixty eighty ninety

- Play the tape and ask the students to check their answers. As they listen they can say the words aloud.

4 Aim: to present the vocabulary in the box.

- Students may already know some of these vocabulary items. You may like to point at certain items in your classroom, such as *cassette player, book, table, chair* etc.

Answers
classroom, self-study room, cassette player, book, table, chair

5 Aim: to present the vocabulary in the box.

- This is the first time in *Reward* Elementary that adjectives have been presented. Illustrate the function of an adjective (a word which gives extra information about a noun) by referring to adjectives in the students' own language.
- Ask the students to decide who or what the adjectives can be used to describe. This activity introduces the concept of collocations.

Suggested answers
person: friendly, beautiful, good, kind, interesting, popular
school: friendly, beautiful, international, good, popular
classroom: beautiful
town: friendly, beautiful, international, interesting, popular
lesson: good, interesting

READING

1 Aim: to practise reading for specific information.

- Explain that a brochure is a document used for purposes of publicising a place or a service.
- Ask the students to read the passage and answer the questions. Explain that they may not understand every word, but tell them that you'll explain four key words, and that they must choose these words carefully.

2 Aim: to check comprehension.

- Ask the students to work in pairs and to check their answers.

Answers
1 Oxford
2 There are two hundred students from thirty countries.
3 Thirty
4 Italy, Brazil, Japan, Greece
5 Andrew
6 (01865) 27462

GRAMMAR

1 Aim: to elicit the rule for using *there is* and *there are*.

- Ask the students to read the information in the grammar box and then to do the exercises.

- Do this exercise orally with the whole class. It's often useful to get the students to work out the rule for themselves rather than simply give it to them.

Answers
1 *is*
2 *are*

2 Aim: to focus on regular plurals.

- Draw the students' attention to the fact that most plurals in English are formed by adding *-s* or *-es* to a noun.

- Ask the students to do the exercise.

Answers
1 computer, rooms
2 tables, chairs, cassette player
3 teachers
4 colleges
5 student

- Tell the students that you use *one* not *a/an* if the actual number is significant.

3 Aim: to focus on the position of adjectives; to focus on word order in sentences.

- Ask the students to put the words in the right order. You may need to explain that *very* is a word which emphasises an adjective.

Answers
1 The classrooms are very comfortable.
2 Oxford is a beautiful city.
3 The class is very international.
4 The lessons are very interesting.
5 Their teacher is friendly.
6 The people in Oxford are kind.

SPEAKING AND WRITING

1 Aim: to practise speaking and writing.

- This is an activity which involves a lot of speaking and a certain amount of writing. Encourage the students to plan their brochures very carefully, and to write them as a collaborative effort.

- You may like to ask the students to do this activity for homework.

2 Aim: to practise speaking; to encourage a response to other people's work.

- If you can, put the various brochures on the wall for everyone to read.

'Oxford is a lovely city. The colleges are beautiful and the people are very kind.' *Nourredine, Morocco*

'The lessons are very interesting.' *Efstathia, Greece*

For more information, here's our address and phone number:

5, George Street, Oxford, England
01865 27462

Or come and visit us. The people at the reception desk are very friendly!

GRAMMAR

There is/are

There's *a door. (= There is a door.)*
There are *seven chairs.* ***Is there*** *a car park?*

Plurals (1): regular

You add *-es* to form the plural of nouns which end in *-s:*
class – classes ***address – addresses***

Position of adjectives

You can put an adjective in two positions:
– after the verb *to be.*
*She**'s kind.*** *Oxford **is lovely.***
– before the noun.
*She's a **kind** woman.* *Oxford is a **lovely** city.*

1 Complete the sentences with *there is* or *there are.*

1 You use ____ when the noun is singular.
2 You use ____ when the noun is plural.

2 Singular or plural?

1 There is a *computer/computers* in the two self-study *room/rooms.*
2 There are *table/tables* and *chair/chairs*, and a cassette *player/players.*
3 There are thirty *teacher/teachers.*
4 The *college/colleges* in Oxford are beautiful.
5 There is one *student/students* from Greece.

3 Put these words in the right order and write sentences.

1 very classrooms are comfortable the
2 Oxford a city is beautiful
3 the is international class very
4 lessons are interesting the very
5 is teacher friendly their
6 the Oxford are kind people in

SPEAKING AND WRITING

1 Work in pairs. Write a brochure for your ideal school. Say:

– what the school is called
– where your school is
– the number of students there are
– the number of teachers there are
– what there is in the school
– what there is in each classroom
– what the address and phone number is

Use the brochure in *Reading* activity 1 to help you.

2 Put your brochures on the classroom wall for other students to read.

5 Where's my pen?

***Has/have got*; prepositions of place (1)**

LISTENING AND VOCABULARY

1 Complete the conversations below with these words.

It's   chair   Where's   Thanks   is   have

A (1) _____ my pen?
B (2) _____ on the table, near your book.
A Oh, I see. Thanks.

A Have you got a mobile phone?
B Yes, I (3) _____. It's in my bag. Here you are.
A (4) _____.

A Where's my bag?
B What colour (5) _____ it?
A It's blue.
B It's under your (6) _____.
A Oh, yes. Thank you.

2 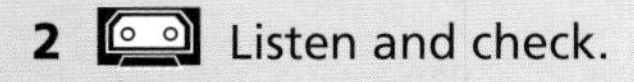 Listen and check.

3 Go round and act out the conversations with other students in the class.

4 Match the words in the box with the items in the photos.

mobile phone   calculator   pen   glasses   watch   camera personal stereo   keys   comb   wallet   diary   ring   bag pencil   book   notebook   coat

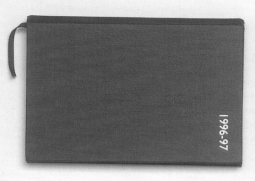

5 Match the words in the box and the colours.

red   orange   yellow   pink   blue   purple   green brown   white   grey   black

6 What colour are the items in the photo?

pink bag ...

SOUNDS

1 Listen and repeat the alphabet.

a b c d e f g h i j k l m
n o p q r s t u v w x y z

2 Put the letters of the alphabet in the correct column.

/eɪ/	/iː/	/e/	/aɪ/	/əʊ/	/uː/	/ɑː/
a	b	f	i	o	q	r

Now listen and check.

GRAMMAR

> *Has/have got*
> **You use *have got* when you talk about possession. *Have got* means the same as *have*. You use it in spoken and informal written English.**
> *He**'s got/has got** a pen. (= He has a pen.)*
> *They**'ve got/have got** a new car. (= They have a new car.)*
> ***Have you got** a watch?   Yes, I have. No, I haven't.*
>
> Prepositions of place (1)
> *He's got a mobile phone **in** his bag. The pen is **on** the table.*
> *The bag is **under** the chair. His pen is **near** his book.*

5

GENERAL COMMENTS

Has/have got

Students sometimes think that because *have got* has the same meaning as *have*, it can always be used to replace it. So it is important to stress that it can only replace the verb *have* when talking about facilities, possession or relationships. It's common to use the contracted form. However, if you use *have* with the same meaning, you don't usually use the contracted form, *I have a new coat.* not *I've a new coat.* You may wish to point out that *have got* is not usually used in the past.

The alphabet

Learning the alphabet is an essential study skill to develop, so you may wish to give the students more practice in this than space in the Student's Book allows.

LISTENING AND VOCABULARY

1 Aim: to prepare for listening; to focus on the target structures.

- Ask the students to read the conversations and to complete them with the words. This pre-listening stage is designed to give them confidence when they listen to the tape.

2 Aim: to check comprehension; to listen for specific information.

- Play the tape and ask the students to check their answers to 1.

Answers
1 Where's
2 It's
3 have
4 Thanks.
5 is
6 chair

3 Aim: to practise using the target structures.

- Act out the conversations with three students in front of the class.
- Ask several students to act out the conversations in pairs in front of the class.
- Ask the students to act out the conversations in pairs. They can change pairs each time.

4 Aim: to present the words in the vocabulary box.

- Ask the students to match the words in the box with the items in the photos.
- Check the students' answers with the whole class.

5 Aim: to present the words in the vocabulary box.

- It is easier to present the words for colours if you have some coloured paper or some magazine photos showing the relevant colours.

6 Aim: to practise using the words presented in this lesson.

- Ask the students to say what colour the different items are in the photo.
- Continue this activity by pointing to objects in the classroom and saying what colour they are.
- Ask the students to continue this activity in pairs.

Answers
pink bag, pink watch, brown comb,
orange notebook, green pencil, black diary,
yellow personal stereo, blue camera, black wallet,
red book, purple glasses, black calculator,
black mobile phone, brown jacket

SOUNDS

1 Aim: to present the pronunciation of the alphabet.

- Write the alphabet on the board. Play the tape and ask the students to listen and repeat.
- Spend some time getting students to practise the pronunciation of any difficult letters, especially any letters which are pronounced differently from the way they're pronounced in the students' own language.
- Point to letters and ask students to say them aloud.
- Ask the students to work in pairs and to take turns in writing and saying the letters.

2 Aim: to practise the pronunciation of the alphabet.

- Say each phoneme aloud.
- Ask the students to put each letter of the alphabet under the relevant phoneme.

Answers
/ei/: a h j k
/iː/: b c d e g p t v
/e/: f l m n s x z
/aɪ/: i y
/əʊ/: o
/uː/: q u w
/ɑː/: r

- Play the tape. Ask students to check their answers.

GRAMMAR

1 Aim: to focus on *has/have got.*

- Ask the students to read the information in the grammar box and then to do the exercises.
- Tell the students two or three things you've got in your bag or your pocket.
- Ask the students to do the same.
- Ask the students to work in pairs and say what they've got in their bags. Once they've done this, they may like to work with different partners.

2 Aim: to practise using *has/have got.*

- Play this game with one or two students in front of the class.
- Ask the students to continue their practise of *has/have got* by playing the game.
- Go round and check that everyone is using the structure correctly.

3 Aim: to focus on prepositions of place.

- Point out that the statements in the grammar box accurately describe the position of the objects in the drawing. Ask the students to decide if the statements in this exercise are true or false.

Answers
All the answers are false.

4 Aim: to practise using the target structures.

- Ask the students to continue to practise using *has/have got* and prepositions of place.

SPEAKING AND LISTENING

1 Aim: to practise speaking; to reactivate passive vocabulary.

- Ask the students to suggest things that typical teenagers have got in their country.
Make a list on the board.
- Ask the students to work in pairs and add four or five items to the list. Ask them to add a comment such as *everyone, most people, a lot of people* by each item.

2 Aim: to practise listening for specific information.

- Make sure everyone understands the meaning of the words for possessions.
- Ask the students to listen and to tick what Steve has got. Play the tape.

3 Aim: to check comprehension; to practise listening.

- Ask the students to work in pairs, trying to remember each of Steve's possessions.
- Ask the students to check their answers. Play the tape again.

Answers
a personal stereo, a computer, a bicycle,
a television, a radio, a watch

4 Aim: to consolidate vocabulary learning.

- This is a memory game, like *Kim's Game*. Ask the students to look at the picture in the Communication activity. Allow them 30 seconds for this.

5 Aim: to practise speaking; to practise using the target structures.

- Ask the students to turn back to page 11, or to close their books.
- Ask the students to describe in detail what they saw and where the objects are. You can do this with the whole class.
- For homework, you may like to ask the students to write a list of what they saw and where it was.

1 Work in pairs. Say two or three things you've got in your bag.

I've got a ...

2 Work with another student.

Student A: In turn, ask Student B what he/she's got in his/her bag. Score 1 point for every *Yes* answer. Answer Student B's questions.

Student B: Answer Student A's questions. In turn, ask Student A what he/she's got in his/her bag.

Have you got a calculator? *Yes, I have.*
Have you got a pencil? *No, I haven't.*

3 Look at the picture. Which of the following statements are true?

1 The bag is on the table.
2 The mobile phone is under the chair.
3 The book is under the table.
4 The pen is in the bag.
5 The calculator is near the glasses.
6 The coat is on the chair.

4 Go round asking and answering:

– what someone has got in their bag
– where something is

What have you got in your bag?
Guido, where's your bag?

SPEAKING AND LISTENING

1 Work in pairs. Make a list of ten possessions a typical teenager has got in your country.

2 Listen to Steve from Britain. Tick (✓) the things that he's got.

radio television bicycle watch personal stereo video mobile phone computer

Did he mention any of the things you wrote in 1?

3 Work in pairs. Try to remember as much as possible.

Now listen again and check.

4 Turn to Communication activity 2 on page 99.

5 Work in pairs. Try to remember what there is in the picture.

Progress check 1–5

VOCABULARY

1 Write new words in a notebook. Write down their meaning in your language and the lesson they come from.

married – marié (Lesson 3), book – livre (Lesson 4)

You can also think of topic headings to put them under.

Jobs – *journalist police officer secretary*
Possessions – *pen watch keys diary*

Here are some words from Lessons 1 to 5.

American age book blue diary doctor
Hungary Korean female red pen
personal stereo taxi driver Thailand first name

Write the words down under one of the topic headings below. (Some words can go under more than one heading.)

countries nationalities jobs classroom colours
personal possessions personal information

Now look back at Lessons 1 to 5 and write down words you want to remember.

2 Here are some questions and replies to use when you have difficulty understanding something.

What's *cartolina* in English? I don't know.
What does *popular* mean? I'm not sure.
How do you say *nom de famille* in English?
I don't understand.

Choose five or six new words from Lessons 1 to 5 which you don't understand, and five or six words in your language which you'd like to know in English.

Now go round the class asking what they mean or what they are in English.

What does 'favourite' mean? I'm not sure.
How do you say 'ami' in English? Friend.

GRAMMAR

1 Put the words in the right order and write sentences.

1 name's Peter my.
2 you married are?
3 are you how?
4 there students how are many?
5 your name what's?
6 pen on the table is.

1 My name's Peter.

2 Write the possessive adjectives which go with these subject pronouns.

I you he she we they

I – my

3 Choose the correct word.

1 *She/her*'s from England.
2 *We/our*'re students.
3 *I/my* age is twenty-two.
4 *He's/his* name is Phil.
5 *Your/you*'re a student.
6 *They/their* coats are on the table.

4 Here are some answers. Write the questions.

1 My name's Andrew.
2 Britain.
3 I'm an artist.
4 Her name's Lah.
5 She's from Bangkok.
6 She's Thai.

1 What's your name?

5 Look at the photo. Write the answers to the questions.

1 What's his name?
2 How old is he?
3 What's his job?
4 Where's he from?

1 His name is Paul Harris.

Name	Paul Harris
Date of birth	20/4/76
Address	London
Occupation	Student

Progress check 1 – 5

GENERAL COMMENTS

You can work through this Progress check in the order shown, or concentrate on areas which have caused difficulty in Lessons 1 - 5. You can also let the students choose the activities they would like or feel the need to do.

VOCABULARY

1 Aim: to organise vocabulary learning.

- Before you come to class, you may like to think about the notes you make or made when you came across new vocabulary. Share this experience and advice with your students.
- It may be a suitable moment to draw your students' attention to the *Wordbank* in their Practice Books.
- Encourage the learners to write the words down under the topic headings.

Possible answers
countries: Hungary, Thailand
nationalities: American, Korean
jobs: doctor, taxi driver
classroom: book, pen
colours: blue, red
personal possessions: diary, personal stereo
personal information: female, first name, age

- Make sure the students understand that some words can go under more than one heading.

2 Aim: to develop communication strategies.

- Write the expressions on the board, and spend time getting the students to learn them by heart. Tell them not to worry about each component. In time, they'll understand how the expression is put together syntactically.
- Ask the students to choose words which they don't know, and to go around asking and saying what the words mean.
- If you haven't done so already, this is a good moment to point out that the target vocabulary in each lesson is contained in the vocabulary box, and that the students should make an effort to ensure they now know these words from Lessons 1 - 5.

GRAMMAR

1 Aim: to revise asking for and giving personal information; to practise word order in sentences.

Answers
1 My name's Peter.
2 Are you married?
3 How are you?
4 How many students are there?
5 What's your name?
6 The pen is on the table.

2 Aim: to revise possessive adjectives and subject pronouns.

Answers
I - my, you - your, he - his, she - her,
it - its, we - our, they - their

3 Aim: to revise possessive adjectives and subject pronouns.

Answers
1 She
2 We
3 My
4 His
5 You
6 Their

4 Aim: to revise questions.

Answers
1 What's your name?
2 Where are you from?
3 What's your job?
4 What's her name?
5 Where's she from?
6 What's her nationality?

5 Aim: to revise giving personal information.

Answers
1 His name is Paul Harris.
2 He is 26 years old in 2002.
3 He's a student.
4 He's from London.

6 Aim: to revise *there is/are.*

- Ask the students to answer the questions with reference to their own classroom.

7 Aim: to revise prepositions of place.

Answers
The coat is on the chair.
The bag is on the table.
The book is on the table.
The glasses are in the bag.
The mobile phone is under the chair.

SOUNDS

1 Aim: to present /iː/ and /ɪ/.

- Play the tape and ask the students to tick and say aloud the words they hear.

Answers
heat sit tick feet rich pip

2 Aim: to revise stress patterns.

- Play the tape. Then ask the students to say the words aloud.

3 Aim: to revise stress patterns.

- Ask the students to match the words with the stress patterns in activity 2.
- Play the tape so the students can check their answers.

Answers
■▪ mother teacher noisy trainers
▪■ police cassette
■▪▪ stereo camera

READING AND WRITING

1 Aim: to practise reading for specific information.

Answers
1 c 2 a 3 b

2 Aim: to present rules of punctuation.

Answers
a You use a capital letter at the start of a sentence, with the first person singular *I*, names of people, countries, nationalities and other proper nouns.
b You use a comma to introduce a new idea in the same sentence, and in long numbers.
c You use a question mark at the end of a question sentence.
d You use a full stop at the end of a statement sentence.

3 Aim: to practise using correct punctuation.

Answers
How many people speak English as a foreign language?
100 million speak it as a foreign language in countries like France, Italy, Brazil and Thailand.
Is there any other language with more speakers?
There are about 1 billion speakers of Chinese as a first language.
How many words do people use in everyday speech?
Most people only use about 10,000 words.

6 Answer these questions with short answers.

1 Is your teacher in the classroom?
2 Are there twenty tables?
3 Are there ten students?
4 Is there a cassette player?
5 Is it a small classroom?
6 Are you a student?

1 Yes, she is.

7 Look at the picture. Write sentences saying where the things are.

The bag is on the table.

SOUNDS

1 Listen and tick (✓) the words you hear.

1 hit heat
2 sit seat
3 tick teak
4 fit feet
5 rich reach
6 pip peep

Now say the words aloud.

2 Listen to the stress pattern in these words.

□ □ England Poland Thailand

□ □ Brazil Japan Peru

□ □ □ Germany Mexico Hungary

Now say the words aloud.

3 Match these words with the stress patterns in 2.

mother teacher police cassette noisy stereo camera trainers

Now listen and check.

READING AND WRITING

1 Match the questions 1 – 3 and the answers a – c about English. Use a dictionary, if necessary.

1 How many people speak English?
2 How many people speak English as a second language?
3 How many words are there in English?

a Another 350 million people, in countries such as Nigeria, Kenya, India and Pakistan.
b There are about 1 million words in English.
c 350 million people speak English as a first language, in countries such as Britain, United States of America, Canada and Australia.

2 Look at the punctuation in the questions and answers in 1. When do you use:

a a capital letter b a comma c a question mark d a full stop

Can you think of other occasions when you use a capital letter in English?

3 Rewrite these three questions and three answers about English with capital letters, commas, question marks and full stops.

how many people speak english as a foreign language 100 million people speak it as a foreign language in countries like france italy brazil and thailand is there any other language with more speakers there are about 1 billion speakers of chinese as a first language how many words do people use in everyday speech most people only use about 10000 words

6 *Families*

Possessive *'s* and *s'*; plurals (2): regular and irregular

VOCABULARY AND SOUNDS

1 Look at the photo of Holly and her family. Can you guess who the people are? Use the words in the box to help you.

brother father mother sister uncle grandfather grandmother aunt husband wife son daughter girlfriend boyfriend

2 Turn to Communication activity 14 on page 102 to check your answers.

3 Work in pairs. Say who the people are in relation to Holly.

Jenny is her mother.

4 Listen to these words. The underlined sound is /ə/.

broth<u>er</u> fath<u>er</u> moth<u>er</u> sist<u>er</u>

5 Underline the other words in the box in 1 with a /ə/ sound.

Listen and check. As you listen, say the words aloud.

GRAMMAR

> Possessive *'s* and *s'*
>
> **You add the possessive *'s* to singular nouns for people instead of *of* to show possession.**
>
> *My husband's name is Philip.*
>
> not ~~*The name of my husband is*~~ ...
>
> **You add ' to plural nouns ending in *-s*.**
>
> *My sons' names are Andrew and Steve.*
>
> not ~~*The names of my sons are*~~ ...
>
> **Remember:**
>
> *My father's name is Jack. (= The name of my father is Jack.)*
>
> *My father's a doctor. (= My father is a doctor.)*
>
> Plurals (2): regular
>
> **For nouns which end in *-y* you drop the *-y* and add *-ies*.**
>
> *family – families baby – babies*
>
> **For other regular plural endings, see Lesson 4 on page 9.**
>
> Irregular
>
> **There are many irregular plurals in English. Here are some of them.**
>
> *child – children man – men woman – women person – people*

1 Look at these sentences. Are they possessive *'s* or contraction *'s*?

1 Holly's mother is Jenny.
2 Andrew's married.
3 Jenny's husband is Philip.
4 Her children's names are Andrew, Steve and Holly.
5 Holly's family is quite big.
6 Antonia's her brother's wife.

6

GENERAL COMMENTS

Apostrophe *'s*

It is possible that your students may become confused with the different uses of the ending *-s* and apostrophe *'s* . You may need to take time during this lesson to point out that:
- the plural of many nouns is formed by adding *s*.
- the apostrophe *'s* when added to *he, she* or *it* is the contracted form of *is*.
- the apostrophe *'s* when followed by *got* is the contracted form of *has*.
- the apostrophe *'s* when added to a noun referring to a person and followed by another noun indicates the possessive *'s* .
- *its* is the possessive adjective and *it's* is the contracted form of *it is*.

The apostrophe is also used to show that a letter is missing in *I'm* = *I am, aren't* = *are not, isn't* = *is not* .

It may be interesting to reflect that the different uses of the apostrophe *'s* ending also cause confusion among native speakers in Britain. There are many examples of the so-called *intrusive apostrophe* used in straightforward plurals in everyday English.
Please have your bag's ready for inspection.
There is some speculation about whether we are witnessing a change in usage which will become standard in the years to come.

VOCABULARY AND SOUNDS

1 Aim: to revise the words in the vocabulary box.

● It is very likely that your students will have come across the words for members of the family at some earlier stage, so this activity is designed as revision. Ask the students to say the words aloud, and to check if they know what they mean.

● Tell the students that all the people are related, and ask them to guess who the people in the photo are using the words in the vocabulary box.

2 Aim: to check comprehension.

● In case the meaning of these words has escaped the students, this activity sends them to the back of the book to check they know the exact meaning.

● Ask the students to work in pairs. They should turn to the Communication activities as instructed and find out who the people in the photo are.

3 Aim: to check comprehension.

● Ask the students to say who the people are in relation to Holly.

Answers
Jenny is her mother.
Andrew is her brother.
Antonia is her brother's wife.
David is her grandfather.
Philip is her father.
Steve is her brother.
Kate is her grandmother.

4 Aim: to practise the pronunciation of /ə/.

● [tape] In the words used to describe the members of the family, there are a large number of unstressed syllables, the /ə/ sound. Play the tape and ask the students to repeat the words. If you are hoping to provide your students with British English as a model for pronunciation, make sure that they do not pronounce the final *r* sound. (American speakers would tend to use the retroflexive *r* at the end of these words.)

5 Aim: to practise the pronunciation of /ə/.

● Ask the students to say the words aloud. Which words have a /ə/sound?

Answers
uncle grandfather grandmother
husband daughter

● Take time to ensure the /ə/ is correctly pronounced.

● [tape] Play the tape and ask the students to say the words aloud.

GRAMMAR

1 Aim: to focus on the distinction between possessive *'s* and contraction *'s*.

● Ask the students to read the information in the grammar box and then to do the exercises.

● It may be a suitable moment to give your students the information provided in *General comments* above.

● To help your students, ask them to identify the part of speech of the words around the *'s*.

Answers
1 possessive
2 contraction
3 possessive
4 possessive
5 possessive
6 contraction, possessive.

2 Aim: to practise placing the apostrophe in the correct position.

● Ask the students to read the passage and put the apostrophe in the right place.

Answers
My parents' names are Michel and Paulette. My mother's a journalist and my father's unemployed. I've got two brothers and a sister. My brothers' names are Pierre and Thierry. My sister's name's Patricia. She's married to Jean. She's a teacher and he's a banker.

3 Aim: to practise using the possessive *'s*.

● Ask the students to explain the meaning of these words using the possessive *'s*.

Possible answers
aunt: my mother's sister or my father's sister
uncle: my mother's brother or my father's brother
cousin: my uncle and aunt's son or daughter
grandfather: my mother's father or my father's father
grandmother: my mother's mother or my father's mother

4 Aim: to practise talking about your family.

● Ask the students to write the names of their family and to ask and say who they are. Write the following on the board.

Andreas is my brother.
Andreas is my brother's name.
My brother's name is Andreas.

● Encourage them to use different expressions as they describe their family.

READING AND SPEAKING

1 Aim: to prepare for reading.

● Ask the students to look at the photo of Kibiri and to decide where he comes from. You may like to spend some time teaching the names of a few countries in Africa. Write the countries on the board.

2 Aim: to prepare for reading.

● The underlined words are those which the students are likely to find difficult. You can ask them to look these up in their dictionaries or you can explain or translate them.

3 Aim: to practise reading for main ideas.

● Ask the students to read the passage and to decide which is Kibiri.

● Kibiri is standing in the middle of the photo. Do not be surprised if your students think that either of the men standing in this photo could be Kibiri. This is a speculative exercise aimed at generating the students' interest in the authentic passage.

Answer
Kibiri is standing in the middle of the photo dressed in white.

4 Aim: to practise reading for specific information; to practise using the target structures.

● This activity is an opportunity for the students to read the passage again and to use the possessive *'s* to describe who the people in the photo are.

5 Aim: to practise speaking; to provide an opportunity for cross-cultural comparison.

● The intention of the passage about Kibiri was to offer a family profile which is probably very different from that of the students' own families. Use this opportunity to explore the differences between Kibiri's family life and your students' own family life. What advantages and disadvantages are there?

● You may like to do this activity in your students' own language, as they may not yet be at a suitable level to do this in English. It may be worth noting at this point that cross-cultural work of this kind can often be done in the students' own language.

2 Put the ' in the right place.

My parents names are Michel and Paulette. My mothers a journalist and my fathers unemployed. Ive got two brothers and a sister. My brothers names are Pierre and Thierry. My sisters names Patricia. Shes married to Jean. Shes a teacher and hes a banker.

3 Explain the meaning of these words.

aunt uncle cousin grandfather grandmother

aunt – my mother's sister or my father's sister

4 Work in groups of four or five. Write a list of your relatives' names. In turn, ask and say who each person is.

READING AND SPEAKING

1 You're going to read about Kibiri and his family. Look at the photo. Where do you think Kibiri lives?

2 Check the meaning of the underlined words in your dictionary.

They <u>live</u> with Kibiri's father.
They all <u>work</u> on the farm.
The women and the girls <u>stay</u> at home with the old and the sick people.
The women <u>make</u> yoghurt and <u>sell</u> it for food.
Kibiri's wife and their daughters <u>walk</u> ten kilometres ...

3 Read the passage and look at the photo. Who is Kibiri?

4 Work in pairs. Look at the photo and say who you think the people are.

5 Work in pairs. Describe a typical family in your country.

In a typical family in my country there are two or three children.

Kibiri is a farmer and a vet in Senegal. He's thirty-five years old. Kibiri and his wife have got seven children, five girls and two boys. They live with Kibiri's father and his brother's family. They all work on the farm. The men and the older boys are often away from home with the animals. The women and girls stay at home with the old and sick people. The women make yoghurt and sell it for food, oil, salt and other things for the home. Kibiri's wife and their daughters walk ten kilometres twice a day to get water. It's hard work but they're a very happy family.

What's the time?

Present simple (3) for customs and routines; prepositions of time (1)

It's one o'clock.

SPEAKING AND VOCABULARY

1 Look at the clocks and listen to the times.

It's five past twelve.

It's twenty to seven.

It's a quarter past one.

It's half past seven.

It's a quarter to five.

Now write the times for the clocks below.

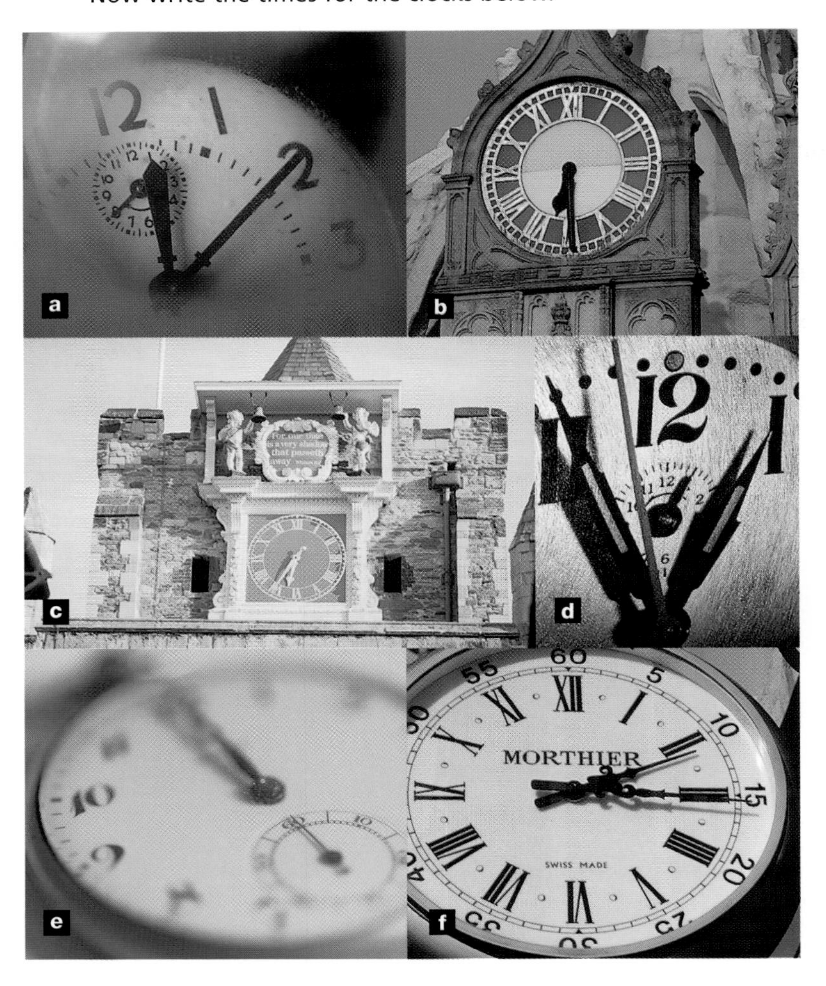

2 Work in pairs and check your answers to 1.

3 Write down three different times.

Now work in pairs. Ask and say what time it is.

Excuse me, what's the time?
It's five to seven.

4 Look at the words in the box. Find:

– five times of the day
– three meals
– two things you do every day

afternoon breakfast dinner evening finish get up go to bed go to work/school lunch midday morning night start weekend work

Complete the sentences with words from the box.

1 You have _____ in the morning.
2 You have dinner in the _____.
3 You _____ at night.
4 People _____ work at five in the afternoon.
5 Many people don't work at the _____.
6 You have _____ at midday or in the afternoon.
7 You get up in the _____.
8 People start _____ at half past eight in the _____.

7

GENERAL COMMENTS

You may like to tell your students that it isn't common to use the twenty-four hour clock in everyday spoken and written English. The am/pm distinction is used if there is any ambiguity.

There may not be enough practice of telling the time in this lesson, and you may like to supplement it with activities from the Practice Book and the Resource Pack.

If you have a cardboard clock for teaching the time, you'll need it for this lesson.

SPEAKING AND VOCABULARY

1 Aim: to present the times of the day.

- Ask the students to look at the clocks and to listen to the tape. You can ask them to repeat the times as they hear them.
- Ask the students to write the times for the other clocks.

Answers
a ten past twelve
b half past six
c twenty-five to seven
d five to one
e twelve o'clock
f a quarter past two

2 Aim: to practise telling the time; to check your answers to 1.

- Ask the students to check their answers to 1.

3 Aim: to practise telling the time.

- Write a few times of the day on the board. Say *Excuse me, what's the time?* Point to a student and elicit a reply.
- Ask the students to work in pairs and do the same.

4 Aim: to present the words in the vocabulary box.

- Ask the students to use their dictionaries if there are any words which they don't understand. You may also like to explain some of the words yourself, although find out if anyone else can do so first.

Answers
five times of the day: afternoon, evening, midday, morning, night
three meals: breakfast, lunch, dinner
two things you do every day: get up, go to bed

- Ask the students to complete the sentences with words from the box.

Answers
1 breakfast
2 evening
3 go to bed
4 finish
5 weekend
6 lunch
7 morning
8 work, morning

READING AND LISTENING

1 Aim: to practise reading for main ideas.

- Ask the students to read the passage about daily routines around the world. Ask them to compare the daily routines with their own.

2 Aim: to practise speaking.

- Ask the students to work in pairs and to compare their answers. It doesn't matter if not all your students come from the same culture. The activity is designed to create an opportunity for cross-cultural comparison.

3 Aim: to practise listening for main ideas.

- [cassette] Ask the students to listen to Tony talking about the statements. They should simply tick the statements which Tony says are true for Australia.

Answers

In Austria children go to school at half past seven in the morning.	✗
In Germany people go to work between seven and nine in the morning.	✓
In Holland people start work at eight in the morning and finish work at five in the afternoon.	✗
In Greece children start school at eight and finish at two pm or start at one pm and finish at six in the evening.	✗
In France people have lunch at midday.	✗
In Spain people have lunch at three or four o'clock in the afternoon.	✗
In the USA people finish work at five in the evening.	✓
In Norway people have dinner at five in the afternoon.	✗
In Spain people have dinner at ten or eleven in the evening.	✗

- Ask the students to check their answers in pairs.

GRAMMAR

1 Aim: to practise using the present simple affirmative and negative.

- Ask the students to read the information in the grammar box and then to do the exercises.

- Ask the students to write sentences which are true for their country.

2 Aim: to focus on prepositions of time.

- The prepositions of time presented in this lesson are complicated by the fact that some of them are followed by the article, and some of them are not.

Answers

1 the 2 no article 3 the 4 no article 5 the

WRITING

1 Aim: to focus on conventions of writing informal letters; to prepare for letter writing.

- Make sure the students understand the differences between informal letters (to friends and family) and formal letters (in business situations, to your bank etc). Make it clear that this letter is an informal one, between two penfriends.

- Ask the students to answer the questions about the layout and other conventions. Make sure they realise that the address is that of the person sending the letter.

Answers

You start an informal letter with *Dear*, followed by the person's first name.
You write your address at the top of the letter on the right-hand side.
You write the date under your address.
You finish it with *Best wishes* and your name.

2 Aim: to practise informal letter writing; to practise using the target structures.

- The letter in 1 is intended to be used as a model for this activity.

- Ask the students to work in pairs and read the letter again. Ask them to brainstorm ideas to put into a similar letter describing daily routines in their own country.

- Ask the students to work on their own and to do a draft of their letters.

- When they are ready, they can look at each other's letters to make suggestions and to borrow ideas.

- Ask the students to do another draft or two of their letters. You may like to set this stage for homework.

- Put the letters on the wall for everyone to read.

READING AND LISTENING

1 Read *Daily routines around the world* and decide which statements are true for your country.

Daily routines around the world

- ☐ In Austria children go to school at half past seven in the morning.
- ☐ In Germany people go to work between seven and nine in the morning.
- ☐ In Holland people start work at eight in the morning and finish work at five in the afternoon.
- ☐ In Greece children start school at eight and finish at one-thirty or start at two and finish at seven in the evening.
- ☐ In France people have lunch at midday.
- ☐ In Spain people have lunch at three or four o'clock in the afternoon.
- ☐ In the USA people finish work at five in the afternoon.
- ☐ In Norway people have dinner at five in the afternoon.
- ☐ In Spain people have dinner at ten or eleven in the evening.

2 Work in pairs and compare your answers.

3 Listen to Tony, from Australia. Tick (✓) the statements in *Daily routines around the world* which are true for Australia.

GRAMMAR

Present simple (3) for customs and routines

You use the present simple to talk about customs and routines.

In Spain people ***have*** *dinner at ten or eleven in the evening.*

The form of the present simple is the same for all persons except the third person singular (*he/she/it*). (For more information on the third person singular form, see Lesson 9.)

I	
You	*leave work at five in the afternoon.*
We	*start work at nine.*
They	

You form the negative with *don't.*

The Australians ***don't*** *have lunch at midday.*

They ***don't*** *have dinner at five.*

Prepositions of time (1)

in: *in the evening   in the morning*

at: *at night   at midday   at midnight   at seven o'clock   at the weekend*

1 Rewrite the statements in *Daily routines around the world* so they are true for your country.

We don't start work at eight. We start at nine.

2 Complete these statements about Australia with *a/an, the* or put a – if no article is needed.

1 In Australia we get up at seven in _____ morning.
2 We have lunch at _____ one o'clock.
3 We stop work at five in _____ afternoon.
4 We go to bed at eleven or twelve at _____ night.
5 We don't work at _____ weekend.

WRITING

1 Read this informal letter to a friend and answer the questions:

– how do you start an informal letter?
– where do you write your address?
– where do you write the date?
– how do you finish a letter?

> 13, Hill Top Road
> Edinburgh
> 12/3/97
>
> Dear Francesca,
>
> Thank you for your letter about the times you do things in Italy. In Britain, we get up at seven or eight o'clock in the morning. We have breakfast at eight o'clock and then we go to work. We work from nine in the morning to five in the afternoon and then we go home. We have dinner at six or seven o'clock in the evening, and we usually go to bed at eleven o'clock or midnight.
>
> Best wishes,
> James

2 Write a letter to a friend. Say what time you do things in your country.

8 Home

Some and _any_ (1); prepositions of place (2)

The Kapralov Family, Sudzal

Number of people in the house: 4
Living area: 140m2
Household equipment: 2 radios, 1 stereo system, 2 telephones, 2 televisions, 1 car
Most treasured possession: video games, Barbie doll
Percentage of income spent on food: 25%

VOCABULARY AND SOUNDS

1 Tick (✓) the rooms and places there are in your home.

sitting room dining room kitchen garden bathroom
bedroom study garage

Now work in pairs. Ask and say what rooms and places there are in your home.

Is there a sitting room?
Yes, there is.

2 Work in pairs. In which rooms do you expect to see the following things?

armchair bath bed bookcase chair cooker
cupboard curtains dishwasher fire fridge lamp
sofa table television toilet shower washbasin

3 Look at these words and underline the stressed syllable.

curtains cooker sofa kitchen study
television shower

4 Work in pairs and check your answers to 3.

Now listen and check.

5 Work in pairs. Look at the box and say where the rooms and places in 1 are in your homes.

upstairs downstairs in the garden outdoors indoors
at the back at the front

Our sitting room is at the back of the flat.

Our dining room is downstairs at the front of the house.

8

GENERAL COMMENTS

Cross-cultural comparison

As the Introduction mentioned, creating the opportunity for cross-cultural comparison is an important part of the syllabus for socio-cultural training. The cultures shown in the *Reward* series are chosen for their interest value as well as for the opportunity they provide to allow the students to reflect on their own cultural background. Furthermore, even with students from the same background, there will be small differences in behaviour, attitudes, customs and beliefs which can be exploited in the classroom for their socio-cultural significance.

VOCABULARY AND SOUNDS

1 Aim: to present the words in the vocabulary box.

- Like the words to describe members of the family, these may also be words which the students already know, so you may not need to explain many of them. Do ask other students to explain difficult words before you explain them yourself.
- Ask the students to work in pairs and to say what rooms there are in their homes.
- Remind the students that they learned the structure *there is/are* in Lesson 4.
- Ask three or four students to tell the whole class which rooms they have in their homes.

2 Aim: to present the words in the vocabulary box.

- Check everyone understands the meaning of the words.
- Ask the students to say in which rooms they would expect to see the words in the box.

Possible answers

sitting room: armchair, bookcase, chair, curtains, fire, lamp, sofa, table, television
dining room: chair, table
kitchen: cooker, cupboard, dishwasher, fridge, table
bathroom: bath, toilet, shower, washbasin
bedroom: bed, curtains
study: bookcase, chair, lamp, table

3 Aim: to focus on syllable stress in words.

- Write the words on the board. Say the words aloud and ask one or two students to underline the stressed syllables.

Answers

curtains cooker sofa kitchen study television shower

4 Aim: to focus on syllable stress in words.

- Play the tape and ask the students to say the words aloud.

5 Aim: to present the words in the vocabulary box.

- Draw a ground plan of a house, and show where the street is. Label the ground plan *at the back, at the front, indoors, outdoors, in the garden*. Draw the front view of the house and label it *upstairs* and *downstairs*.
- Ask the students to say where the rooms in 1 are in their homes.

READING

Aim: to practise reading for specific information.

● Ask the students to read the information about the Kapralov family and look at the photo to decide if the statements are true or false.

Answers
1 True
2 False
3 True
4 False
5 True
6 False

GRAMMAR

1 Aim: to focus on the use of *some* and *any*.

● Ask the students to read the information in the grammar box and then to do the exercises.

● Do this activity orally with the whole class.

Answers

1 any	2 some	3 some
4 any	5 any	6 some

2 Aim: to focus on the difference between *they're* and *their*.

● Remind the students that:
their is a possessive adjective
they're is the contracted form of *they are*.
They are both pronounced in the same way as *there*.

Answers
1 They're 2 Their 3 They're
4 There 5 Their 6 There

3 Aim: to present prepositions of place.

● Ask students to look at the photo and see if they can say where things are. Remind them to use the prepositions *in front of* and *behind* in the instances where they can't see the various items.

Answers
1 There's a chair next to the television.
2 There's a clock next to the sofa.
3 There's a car behind the bikes.
4 There's a chair next to the table.
5 There are two armchairs behind the carpet.
6 There's a bookcase behind the sofa.

LISTENING AND WRITING

1 Aim: to practise listening for specific information.

● Ask the students to guess Geoff's answers to the statements in the chart on the tape.

● Play the tape and ask the students to listen and check.

2 Aim: to check comprehension.

● Ask the students to work in pairs and to complete the chart.

Answers
Type of home: boat
Size: 10 metres long and 2 metres wide
Number of rooms: 5
Number of people: 3
Furniture: fridge, cooker, several armchairs, a television, a shower
Most important item: computer

● Play the tape and ask the students to check their answers.

3 Aim: to practise speaking.

● Ask the students to talk about their own homes and to complete the chart.

● You may like to ask them to write a paragraph about their partner's home for homework.

READING

Read the information about the Kapralov family. Are these statements true or false?

1 There are some curtains.
2 There aren't any carpets.
3 They've got some video games.
4 They haven't got any bookcases.
5 There are some cupboards in the kitchen.
6 They haven't got any mirrors.

GRAMMAR

Some and *any* (1)

You usually use *some* with plural nouns in affirmative sentences when you're not interested in the exact number.

*There are **some** chairs in the sitting room.*

You usually use *any* with plural nouns in negative sentences and questions.

*There aren't **any** curtains in the bathroom.*
*Are there **any** cupboards in the kitchen?*

Prepositions of place (2)

*The car is **in front of** the bookcase.*
*The bookcase is **behind** the car.*
*The radio is **next to** the television.*

(For more information on prepositions of place, see Lesson 5.)

1 Complete the sentences with *some* or *any*.

1 There aren't _____ bookcases in the dining room.
2 They have _____ curtains in the sitting room.
3 There are _____ flowers on the table.
4 They haven't got _____ lamps in the kitchen.
5 Are there _____ carpets in the house?
6 They've got a table and _____ chairs in the study.

2 Choose the correct word.

1 *There/they're* from London.
2 *Their/there* children are at school.
3 *They're/their* not at home.
4 *There/they're* aren't any mirrors.
5 *Their/they're* house is quite large.
6 *There/their* are two doors.

3 Look at the photo of the Kapralov family and say where these things are.

1 bookcase/television
2 plant/lamp
3 window/door
4 chair/table
5 armchair/fire
6 mirror/door

There's a bookcase behind the television.

LISTENING AND WRITING

1 You're going to hear an interview with Geoff, from Scotland, who lives on a boat. Look at the photo. Can you guess the answers to the questions in the chart?

	Geoff	Your partner
Type of home	*boat*	
Size		
Number of rooms		
Number of people		
Furniture		
Most important items		

Listen and check.

2 Work in pairs and complete the chart.

Now listen again and check.

3 Find out about your partner's home. Complete the chart.

9 How do you relax?

Present simple (4) for habits and routines:
Wh-* questions; third person singular *(he/she/it)

VOCABULARY AND READING

1 Match the phrases in the box with the pictures below.

drink coffee eat an apple go running listen to the radio play football read a newspaper watch a tennis match have a shower learn the guitar

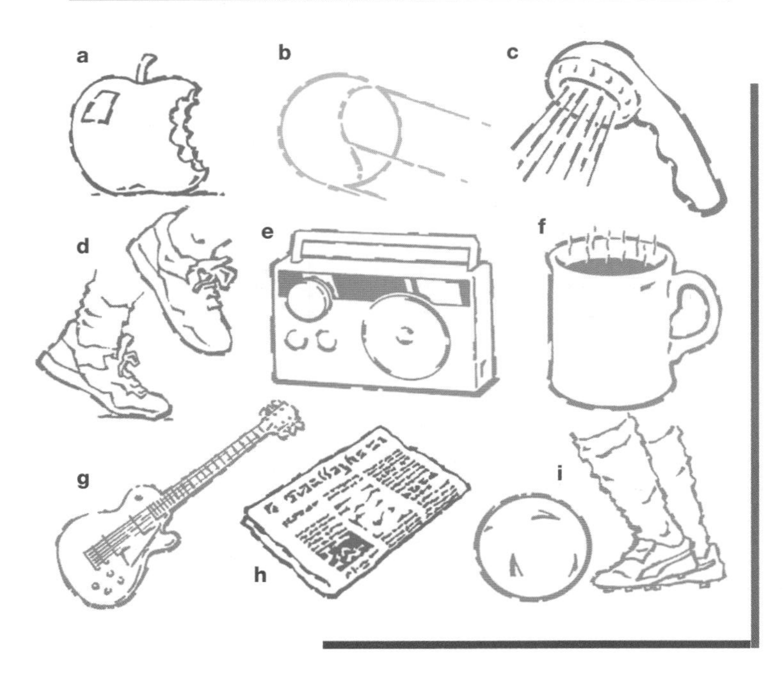

2 Read *How do you relax?* Say who you can see in the photos.

3 Answer the questions.

1 Where does Fiona live?
2 What does Newton do?
3 When does Ingrid go running?
4 Why does Otto get a video?
5 How does Patricia relax?
6 Who does Tanya live with?

4 Write down the verbs in the vocabulary box in 1.

drink, eat ...

Now find phrases in the passage which go with the verbs.

drink tea, eat toast, chocolate ...

How do you relax?

'I go running in the evening, and then I come home and have a shower.' **Ingrid**, Tromso

'My husband goes to his club to relax, and I sometimes go with him.' **Tanya**, Marseille

'I have a hot bath, and listen to the radio.' **Patricia**, Brussels

'I play the guitar, mostly *Rolling Stones* songs.' **Karl Heinz**, Vienna

'My girlfriend likes television, so I get a video and we watch it together.' **Otto**, Budapest

'I drink tea and eat toast and jam in the garden.' **Carrie**, Manchester

'I eat chocolates while my boyfriend reads novels to me.' **Fiona**, Hong Kong

'I learn languages to relax. I'm learning Russian at the moment. My wife learns languages too.' **Gerard**, Lourenco Marques

'I never relax. I'm a taxi driver.' **Newton**, Rio

9

GENERAL COMMENTS

Listening

Like the reading passages in *Reward* Elementary, the listening passages contain a mixture of very carefully controlled and graded language and roughly graded material, which may be slightly above the level which the student might expect. Listening comprehension is generally the skill which the students find most hard to acquire. Unlike reading, they are not in control of the rate of delivery when listening in a foreign language, and practice in this skill often generates a certain amount of panic and confusion. The students need to be reminded that in real life, they won't be able to understand every word of spoken English, and that it is a good idea to expose them to these circumstances in the reassuring and supportive context of the classroom.

VOCABULARY AND READING

1 Aim: to present the words and expressions in the vocabulary box.

● Ask the students to match the words in the box with the drawings.

Answers
a eat an apple
b watch a tennis match
c have a shower
d go running
e listen to the radio
f drink coffee
g learn the guitar
h read a newspaper
i play football

2 Aim: to practise reading for main ideas.

● Ask three or four students to say how they relax. Encourage them to use words and expressions from the vocabulary box in 1.

● Ask the students to read the passage and to say who the people in the photos are.

Answers
Carrie, Manchester
Tanya's husband, Marseille

3 Aim: to check comprehension; to present the target structures.

● Ask the students to read the passage again and to find the answers to the questions.

● You may like to check the students' answers to the questions with the whole class.

Answers
1 In Hong Kong.
2 He's a taxi driver.
3 In the evening.
4 Because his girlfriend likes television.
5 In a hot bath, listening to the radio.
6 Her husband.

4 Aim: to practise using the words in the vocabulary box.

● Explain that the verbs in the expressions in the vocabulary box can be used with many other expressions. These are known as collocations. For example, the verb *drink* collocates with *tea, coffee, water, wine, juice* etc.

● Ask the students to read the passage and find other collocations for the verbs.

Answers
drink tea
eat toast and jam, chocolates
go to a club
listen to the radio
play the guitar
read novels
watch a video, television
have a shower, a hot bath
learn a language

● Ask the students to suggest other collocations they can think of for these verbs.

GRAMMAR

1 Aim: to focus on the form of the third person singular.

● Ask the students to read the information in the grammar box and then to do the exercises.

● Tell the students that the passage is mostly in the first person singular, but that there are some examples of the third person. Remind them to look for verbs ending in *-s* or *-es*.

Answers
goes, likes, reads, learns

2 Aim: to practise the third person singular ending.

● Ask the students to transform the first person singular verb endings into third person singular.

Answers
Ingrid goes running.
Tanya sometimes goes with her husband to his club.
Patricia has a hot bath and listens to the radio.
Karl Heinz plays the guitar.
Otto watches a video with his girlfriend.
Carrie drinks tea and eats toast and jam in the garden.
Fiona eats chocolates while her boyfriend reads novels to her.
Gerard learns languages.
Newton doesn't relax.

3 Aim: to focus on word order in question forms and the use of the auxiliary.

● The use of *do/does* in questions may cause students some difficulties, especially for those whose language uses a simple inversion to form a question. Explain that the use of the auxiliary verb in English is not optional. It has to be used when you ask a *wh-* question.

● Ask the students to complete the sentences.

Answers
1 does, do 2 go 3 What, like
4 does, watch 5 Where, eat 6 does, relax

4 Aim: to practise using the third person singular.

● Ask the students to work in pairs and ask and answer the questions in 3.

Answers
1 She goes running.
2 He goes to his club.
3 He likes the *Rolling Stones*.
4 He watches a video with his girlfriend.
5 She drinks tea and eats toast in the garden.
6 Because he's a taxi driver.

5 Aim: to practise speaking; to practise using the present simple for habits and routines.

● Ask the students to talk in pairs about what they like doing to relax.

SOUNDS

1 Aim: to focus on the pronunciation of third person singular endings.

● Ask the students to listen as you say the three verbs aloud. Emphasise the endings.

● Play the tape and ask the students to repeat the verbs.

2 Aim: to focus on the pronunciation of third person singular endings.

● Write the three phonemes on the board. Ask the students to say the verbs aloud, and then ask students to come up and write the verb under the correct phoneme.

● Play the tape. The students check their answers.

Answers
/s/: makes, eats, drinks, likes
/z/: lives, listens, plays, reads
/ɪz/: finishes, washes

3 Aim: to focus on the *he* and *she* sounds.

● Play the tape and ask the students to listen and tick the sentences they hear.

Answers
1a 2b 3b 4a

LISTENING AND SPEAKING

1 Aim: to prepare for listening.

● This activity can also be used to pre-teach any difficult items.

● Ask the students to tick the things they do in their free time.

2 Aim: to practise listening for main ideas.

● Play the tape and ask the students to tick the things Helen and Chris do.

	Helen	Chris
go to the cinema		
watch television		✓
listen to the radio		
do some sport		
learn a language	✓	
play music		
read a novel	✓	
see friends		✓
go to a club		✓

3 Aim: to practise the target structures.

● Ask the students to check their answers on the chart. Make sure they are formulating third person singular sentences correctly.

4 Aim: to check comprehension.

● Play the tape again and ask the students to check their answers.

● You may like to ask the students to write a few sentences for homework, describing what Helen and Chris do to relax.

GRAMMAR

Present simple (4) for habits and routines

You use the present simple to talk about habits and routines. We usually say *how often* or *when* we do something when describing a routine.

*I **go** running in the evening and then I **come** home and have a shower.*

***Wh-* questions**

You form *Wh-* questions for all verbs except *to be* with *do* or *does* + infinitive. (For *Wh-* questions with *to be*, see Lesson 3.)

***What do** you **do** to relax? **Where does** Fiona **live**?*

Third person singular (*he/she/it*)

You form the third person singular by adding *-s* to most verbs.

drink – drinks   eat – eats   learn – learns
like – likes

You add *-es* to *do, go* and verbs which end in *-ch, -ss, -sh* and *-x*.

do – does   go – goes   wash – washes

(For more information, see Grammar review page 107.)

1 Find four examples of the third person singular in the passage.

2 Work in pairs. Read *How do you relax?* again. Ask and say what the people do to relax.

What does Ingrid do to relax?
She goes running.

3 Complete the questions about the passage.

1 What _____ Ingrid _____ in the evening?
2 Where does Tanya's husband _____?
3 _____ music does Karl Heinz _____?
4 Who _____ Otto _____ a video with?
5 _____ does Carrie drink tea and _____ toast?
6 Why _____ Newton never _____?

Now find the answers.

4 Work in pairs. Ask and answer the questions in 3.

5 Work in pairs. Ask and say what you do to relax.

SOUNDS

1 Listen and repeat these verbs.

/s/	/z/	/ɪz/
gets	does	watches

2 Look at these verbs and put them in the correct columns above.

lives makes finishes washes eats drinks listens plays reads likes

Now listen and check.

3 Listen and tick (✓) the sentences you hear. Do you hear *he* or *she*?

1 a What does he do? b What does she do?
2 a Where does she live? b Where does he live?
3 a How does he see it? b How does she see it?
4 a Where does she go? b Where does he go?

LISTENING AND SPEAKING

1 Tick (✓) the things you do in your free time or when you relax.

	You	Helen	Chris
go to the cinema			
watch television			
listen to the radio			
do some sport			
learn a language			
play music			
read a novel			
see friends			
go to a club			

2 Listen to Helen and Chris. Tick (✓) the things they do in their free time.

3 Work in pairs and check your answers to 2.

Helen sees friends and goes to the cinema.

4 Listen again and check.

10 Do you like jazz?

Pronouns; present simple (5): talking about likes and dislikes

VOCABULARY AND READING

1 Tick (✓) the things in the box that you like.

football cooking tennis magazines going to parties swimming classical music dancing jazz going to the theatre going to the cinema watching sport rock music Chinese food

2 Put the words and phrases in 1 in two columns: *nouns* and *words or phrases ending in -ing*.

nouns: football *words or phrases ending in -ing: cooking*

3 Read the adverts and find an event for someone who likes:

jazz music Chinese food ballet films magazines

4 Write down information about the events you chose in 3.

Name					
Type					
Date/days					
Time					
Place					
Other details					

LISTENING

1 Listen to the conversation. Underline anything which is different from what you hear.

A Do you like classical music?
B No, I don't. I hate it.
A What kind of music do you like? Do you like rock music?
B Yes, I do. I love it.
A Who's your favourite singer?
B Mick Hucknell from *Simply Red*. He's great.
A I don't like him very much. I don't like *Simply Red*.
B Oh, I like them very much. What about you? What's your favourite music?
A I like jazz. My favourite musician is Stan Getz. Do you like jazz?
B It's all right.

2 Work in pairs and correct the conversation in 1.

Now listen and check.

3 Work in pairs and act out the conversation.

10

GENERAL COMMENTS

Like + -ing or noun

This lesson briefly presents the language for talking about likes and dislikes. You may like to explain that the *-ing* form of verbs operates just like a noun phrase. In *Reward* Pre-intermediate these structures will be further explored.

Topicality

It is impossible for a textbook such as *Reward* to remain topical for the whole of its life. It's important, therefore, to see this lesson as a blueprint for discussion, and to replace any outmoded references with more topical ones.

VOCABULARY AND READING

1 Aim: to present the words in the vocabulary box.

- This lesson overlaps in subject matter with Lesson 9. Some of the expressions learnt in that lesson can also be used in this one, so take every opportunity for revision.

- Check that everyone understands the words in the vocabulary box. Ask other students to explain any difficult words, if necessary.

2 Aim: to focus on the difference between nouns and *-ing* form words.

- Ask the students to put the words in two columns.

Answers

noun: football, tennis, magazines, classical music, jazz, rock music, Chinese food

***-ing* words and phrases:** cooking, going to parties, swimming, dancing, going to the theatre, going to the cinema, watching sport

- Ask the students to think of other expressions using the same words. For example, *watching football, watching tennis, Italian food.*

- Make a list of all the possibilities on the board.

3 Aim: to practise reading for specific information.

- Explain to the students that they will probably not know every single word they see printed in the adverts, but they should scan the texts to find events for the people described.

Answers

jazz music – Gill Scot-Heron
Chinese food – The Lotus Flower
ballet – London Festival Ballet
films – Warner Brothers cinema
magazines – Time Off

4 Aim: to practise reading for specific information.

- Ask the students to complete the chart. There may not be information given for all the events.

Name	London Festival Ballet	G.I Scot-Heron	Warner Brothers Multiplex	Time Off	The Lotus Flower
Type	Ballet	Jazz	Cinema	Magazine	Chinese Restaurant
Date days	26th/27th August	12th/13th/14th September	-	-	-
Time	-	-	-	-	-
Place -	Coliseum	Jazz Café, Camden Town			- -
Other details	Swan Lake	-	Twister, Kids, Richard III, The Hunchback of Notre Dame	Tells you how to spend your freetime	Suggests people try Dim Sum and Peking Duck

LISTENING

1 Aim: to practise listening for specific information.

- This activity is intended to make the students listen very carefully to the conversation. It is hoped that the language in the conversation is graded to the students' present level of comprehension.

- Play the tape and ask the students to underline anything which is different from what they hear.

Answers

A Do you like classical music?
B No, I don't. I hate it.
A What kind of music do you like? Do you like rock music?
B Yes, I do. I love it.
A Who's your favourite singer?
B Mick Hucknall from *Simply Red.* He's great.
A I don't like him very much. I don't like *Simply Red*.
B Oh, I like them very much. What about you? What's your favourite music?
A I like jazz. My favourite musician is Stan Getz. Do you like jazz?
B It's all right.

2 Aim: to check comprehension.

- Ask the students to correct the written conversation with what they heard.

- Play the tape so students can check their answers.

3 Aim: to practise speaking.

- Ask the students to act out the conversation. When they have done it two or three times with different students, ask two or three pairs to act it for the whole class.

GRAMMAR AND FUNCTIONS

1 Aim: to present subject and object pronouns.

- Ask the students to read the information in the grammar box and then to do the exercises.
- Ask the students to look at the grammar box and to count the number of pronouns which have the same form as subject and object pronouns.

Answer
The following pronouns are the same as subject and object: *you, it*

2 Aim: to focus on the language for talking about likes and dislikes.

- Ask the students to complete the conversation. You may like to do this orally with the whole class.

Answers
A Do you like Demi Moore?
B Yes, I do. She's great. What about you? Do you like her?
A No, I don't. I don't like her at all.
B Who's your favourite film star?
A I like Jodie Foster.

- Ask the students to work in pairs and to act out the conversation once as it stands and once replacing the names with their own favourite actors and actresses.

3 Aim: to practise using subject and object pronouns.

- Ask the students to do this activity orally in pairs.

Answers
1 her 2 them 3 we 4 him

4 Aim: to practise using the target structures; to prepare for speaking in activity 5.

- Ask the students to think about their favourite well-known people and places. They should make a list.

5 Aim: to practise using the target structures.

- Ask the students to ask and say who their favourite people and places are.
- Find out who are the most popular people and what are the most popular places amongst the students.

SOUNDS

1 Aim: to focus on the tone of the voice.

- Point out to your students that effective pronunciation does not simply involve the correct pronunciation of individual sounds. It also involves the tone of voice. British English is relatively discrete with little of the tone variation that some languages have. But it nevertheless has the ability to express the speaker's relative interest in what he or she is saying or responding to.
- Play the tape. Ask the students to listen and decide who sounds more interested.

Answer
Speaker B

2 Aim: to practise using an interested tone.

- Ask the students to act out the dialogue in pairs. They should both try to sound interested.

LISTENING AND WRITING

1 Aim: to prepare for listening.

- Ask the students to look at the penfriend adverts and to see if there is anyone who has similar interests.

2 Aim: to practise listening for main ideas.

- Play the tape. Ask the students to listen to four people and to focus on what they like doing.

Answer
John: sport - football, tennis, skiing, rock music
Kate: sport - skiing, ice-skating, going to the cinema
Keith: computer gameS, water sports, swimming, water polo, dancing
Susie: going to the beach, dancing, swimming, reading novels, going to the theatre

- Ask the students to decide on a suitable penfriend for Octavio, Tomasz and Nancy.

Possible answers
Keith and Octavio
Kate and Tomasz
John and Nancy

3 Aim: to practise writing.

- Ask the students to write similar paragraphs about themselves, modelled on the adverts.
- Put the adverts on the wall. Are there any possible penfriends to be matched in your class?

GRAMMAR AND FUNCTIONS

> **Pronouns**
>
> ***I*** *like Simply Red. I like* ***them*** *very much.*
> *He likes Stan Getz but I don't like* ***him*** *at all.*
>
> **Subject:** ***I you he she it we they***
> **Object:** ***me you him her it us them***
>
> **Present simple (5): talking about likes and dislikes**
>
> *What kind of music do you like?* *I like classical.*
> *I don't like rock.*
>
> *Do you like jazz?* *Yes, I do. I love it.*
> *Yes, it's great. It's all right.*
> *No, I don't like it very much.*
> *No, I hate it.*

1 Look at the pronouns in the grammar box. How many object pronouns have the same form as their subject pronouns?

2 Complete the sentences.

A Do you _____ the actress Demi Moore?
B Yes, I do. _____ great. What about you? Do you like her?
A No, I _____. I don't like _____ at all.
B Who's your _____ film star?
A _____ like Jodie Foster.

TIMEOFF
TIMEOFF
The Magazine for Entertainment
Get down, get down!
The magazine which tells you all you need to know about how to spend your free time!
For the authentic taste of China, eat at
The Lotus Flower
Try our tasty Dim Sum and Peking Duck

3 Choose the correct pronoun.

1 John likes Sharon Stone but Patti doesn't like *she/her* at all.
2 My sister hates Stone Roses but I like *they/them* a lot.
3 *We/us* both like jazz.
4 My favourite singer is George Michael, but my brother doesn't like *he/him* at all.

4 Write down the names of your favourite:

film star singer sportsperson politician
club TV programme film restaurant

5 Find out if other students like the people or things you wrote in 4.

SOUNDS

1 Listen to the conversation in *Grammar and functions* activity 2. Which speaker sounds interested?

2 Work in pairs and act out the conversation. Try to sound interested.

LISTENING AND WRITING

1 Look at the adverts for penfriends. Is there anyone who likes the same things as you?

> **My name's Octavio** and I'm from Recife in Brazil. I like playing football on the beach, dancing and playing computer games. Write to me.
>
> **I'm Tomasz** and I'm from Wroclaw in Poland. I like ice skating, going to the cinema and photography. I also like watching sport. What about you?
>
> **I'm Nancy** and I live in Seattle. I'm a high school student and my favourite group is *REM*. I really like the singer. I also like playing volleyball, going to the theatre and skiing. Tell me about yourself.

2 Listen and choose a penfriend for Octavio, Tomasz and Nancy.

3 Write an advert for yourself. Use the adverts in 1 to help you. Say:

– who you are – where you're from – what you like doing

Progress check 6–10

VOCABULARY

1 When you write a new word in your notebook, check you know its part of speech. Here are some of them.

noun (n) – *brother father girlfriend film*
verb (v) – *have finish start*
adjective (adj) – *favourite popular comfortable*
pronoun (pr) – *he she it*
preposition (prep) – *in under*

What parts of speech are the following words?

breakfast from husband on read football restaurant to sitting room start telephone red beautiful

2 It's also useful to write down words and expressions which go with verbs.

have	a shower, breakfast
go	to the cinema, to work
play	tennis, football
listen to	music, the radio
watch	football, the television
leave	work, school

Match the verbs on the left with these words and expressions. There may be more than one possibility.

to bed dinner a video a concert the guitar tennis

3 A good dictionary helps you learn more about a new word. Look at this dictionary extract and find out which of the following it shows:

meaning pronunciation word stress part of speech how the word is used

breakfast /ˈbrekfəst/ *n* breakfast is the first meal of the day which most people eat early in the morning. EG I get up at eight o'clock and have breakfast.

Look at your own dictionary and find out what features it shows.

GRAMMAR

1 Write the plural form of these words.

1 child 2 boy 3 man 4 woman
5 brother 6 family

1 children

2 Put the ' in the right place.

Ive got two brothers. My brothers names are Tom and Henry. Toms a doctor and Henrys a teacher. Toms married and his wifes name is Jean. Theyve got two children. Henry isnt married but hes got a girlfriend. Her names Philippa.

I've got two brothers.

3 Put these words in the right order to make sentences.

1 in shops Britain at close five-thirty.
2 get up seven I o'clock at.
3 Greece start at eight children in school.
4 the at USA start people work in nine.
5 Holland in have at they dinner six.
6 we the evening go bed at eleven in to.

1 In Britain shops close at five-thirty.

4 Write sentences saying what time you do these things.

1 get up
2 have breakfast
3 leave home
4 have lunch
5 leave work/school
6 have dinner

1 I get up at seven in the morning.

5 Write the third person singular of the present simple tense of these verbs.

1 do 2 like 3 go 4 make 5 play 6 eat
7 watch 8 learn

1 do – does

Progress check 6 – 10

GENERAL COMMENTS

You can work through this Progress check in the order shown, or concentrate on areas which have caused difficulty in Lessons 1 - 5. You can also let the students choose the activities they would like or feel the need to do.

VOCABULARY

1 Aim: to present the parts of speech.

● Use the students' own language, if you can, to illustrate what a noun, verb, adjective, pronoun and preposition are. Does their language have similar words or are some parts of speech contained within suffixes, inflexions or in some other way?

● Ask the students to decide what part of speech the words are.

Answers
breakfast (n), from (prep), husband (n), on (prep), read (v), football (n), restaurant (n), to (prep), sitting room (n), start (v), telephone (n), red (adj), beautiful (adj)

2 Aim: to focus on collocations.

● Remind the students that a collocation is when two words go together. They did some work on this in Lesson 9.

Possible answers

go to bed	have dinner
watch a video	listen to a concert
play the guitar	play tennis

3 Aim: to present ways of using a dictionary.

● You may or may not have access to English dictionaries in your class. Even if you haven't, it's useful to point out to your students that a dictionary is an important tool in the language learning process, and is nearly as important as the coursebook itself. Encourage them to think about obtaining a copy of a dictionary such as the Cobuild dictionary.

● If you can bring in a few copies of dictionaries for your students for this lesson, this activity will provide a useful introduction to the use of dictionaries in the classroom.

GRAMMAR

1 Aim: to revise plural forms.

Answers
1 children
2 boys
3 men
4 women
5 brothers
6 families

2 Aim: to revise the use of the apostrophe.

Answers
I've got two brothers. My brothers' names are Tom and Henry. Tom's a doctor and Henry's a teacher. Tom's married and his wife's name is Jean. They've got two children. Henry isn't married but he's got a girlfriend. Her name's Philippa.

3 Aim: to revise word order.

Answers
1 In Britain shops close at five-thirty.
2 I get up at seven o'clock.
3 In Greece children start school at eight.
4 In the USA people start work at nine.
5 In Holland they have dinner at six.
6 We go to bed at eleven in the evening.

4 Aim: to revise telling the time and talking about daily routines.

● Ask the students to answer the questions describing their own daily routines.

5 Aim: to revise third person singular endings.

Answers
1 does
2 likes
3 goes
4 makes
5 plays
6 eats
7 watches
8 learns

6 Aim: to revise *there is/are.*

Answers
1 Yes, there is.
2 Yes, there are.
3 Yes, there is.
4 Yes, there are.
5 No, there isn't.
6 Yes, there is.

7 Aim: to revise *some* and *any*.

Answers
1 some
2 any
3 some
4 any
5 any
6 some

8 Aim: to revise subject and object pronouns.

Answers
1 it 2 We 3 her 4 I 5 them 6 me

SOUNDS

1 Aim: to practise the /ə/ sound.

● Play the tape. Ask the students to listen and underline the /ə/ sound.

Answers
mother cousin dinner listen
television computer newspaper

● Ask the students to repeat the words.

2 Aim: to practise the /s/ and /z/ endings for plurals.

● Play the tape so students can check their answers.

Answers
/s/: drinks, Greeks, journalists, desks, sports
/z/: computers, pens, sons, schools

3 Aim: to focus on the tone of voice.

● Say one or two of the sentences in an interested, lively way, and then in a dull, bored-sounding way.

● Ask one or two students to read out the sentences in an interested and a bored way.

● Play the tape and ask the students to say if the speaker sounds interested or bored.

Answers
1 interested 2 bored 3 interested
4 bored 5 interested 6 bored

READING AND SPEAKING

1 Aim: to practise reading for main ideas; to compare cultures.

● The passage describes some life events and at what age they happen in Britain. Once again, this is an opportunity for the students to compare their cultures with another.

● Ask them to decide which ones are true for their country. Are there any which surprise them?

2 Aim: to practise speaking; to compare cultures.

● Ask the students to work in pairs and to talk about the age they do things in their countries.

● You may like to ask one or two pairs of students to talk about the age they do things in front of the whole class.

6 Look at the picture and give short answers to these questions.

1 Is there a television?
2 Are there any curtains?
3 Is there a table?
4 Are there any chairs?
5 Is there a lamp?
6 Are there any bookcases?

1 Yes, there is.

7 Complete the sentences with *some* or *any*.

1 There are _____ curtains in the bedroom.
2 There aren't _____ carpets.
3 There are _____ chairs in the living room.
4 Are there _____ armchairs?
5 They haven't got _____ desks.
6 They've got _____ trees in the garden.

8 Complete these sentences with subject or object pronouns.

1 He likes football but I don't like _____ at all.
2 _____ love tennis and so do our children.
3 Jane's favourite film star is Demi Moore, but I don't like _____ very much.
4 My friend Geoff likes classical music and so do _____.
5 My favourite singers are Pavarotti and Domingo, but Jane doesn't like _____.
6 I like her but she doesn't like _____.

SOUNDS

1 Listen and underline the /ə/ sound in these words.

mother cousin dinner listen television computer newspaper

Now say the words aloud.

2 Say these words aloud. Is the underlined sound /s/ or /z/?

drinks Greeks journalists desks computers pens sons schools sports

Listen and check.

3 Listen to these questions. Put a tick (✓) if the speaker sounds interested.

1 What's your name?
2 What's your job?
3 Where's your mother from?
4 What music do you like?
5 What's your favourite group?
6 What time do you get home from work?

Say the questions aloud. Try to sound interested.

READING AND SPEAKING

1 Read the facts about *The age you do things in Britain.* Which ones are true for your country?

The age you do things in Britain

- ☐ Children start school at five.
- ☐ Children leave school when they're sixteen or eighteen.
- ☐ People learn to drive when they're seventeen.
- ☐ Students go to university when they're eighteen or nineteen.
- ☐ People leave university when they're twenty-one or twenty-two.
- ☐ People get married between the age of twenty and thirty.
- ☐ Men stop work when they're sixty-five.

2 Work in pairs. Talk about the age you do things in your country.

In my country, children start school at six.

11 A day in my life

Present simple (6): saying how often you do things; prepositions of time (2)

READING

1 You're going to read a magazine article about Tanya Philips, who is a presenter on breakfast television in Britain. What time do you think her day starts and finishes?

2 Read the article and find out what time her day starts and finishes.

3 Decide where these sentences go in the article.

a After the programme we always have breakfast and relax.
b I often go shopping and have lunch with friends.
c We usually have dinner quite early, at seven o'clock.

4 Write the questions the journalist asked Tanya.

What time do you get up?

GRAMMAR

> Present simple (6): saying how often you do things
>
> **You use the present simple and the following adverbs of frequency to say how often you do things.**
>
> *100% ______________________ 0%*
> *always usually often sometimes never*
>
> **You usually put the adverb before the verb.**
>
> *I **always** get up at seven o'clock.*
> *I **sometimes** go shopping in the evening.*
> *I **often** have a drink with friends.*
> *I **never** do the washing up.*
>
> **But you put the adverb after the verb *to be*.**
>
> ***I'm usually** asleep by nine o'clock.*
>
> Prepositions of time (2)
>
> ***at:** at seven o'clock at half past three at the weekend*
> ***on:** on Sunday on Tuesday on Monday morning*
> ***from ... to:** from Monday to Friday from seven to nine o'clock.*

My working day starts very early. From Monday to Friday I get up at half past three and I have a shower and a cup of coffee. I usually leave the house at ten past four because the car always arrives a few minutes early. I get to the studio at about five o'clock and start work. *Good Morning Britain* starts at seven o'clock and finishes at nine o'clock. Then I leave the studio at a quarter past ten. After that, I get home at three o'clock. A woman helps me with the housework and the ironing. I read the newspaper and do some work. Then my husband gets home at half past five in the afternoon and I cook dinner. We stay at home in the evening. We don't go out because I go to bed very early. We usually watch television and then I go to bed at half past eight. I'm usually asleep by nine o'clock. At weekends I don't get up until ten o'clock. In the evening, we often see some friends or go to the cinema. But I'm always up early again on Monday morning.

11
Adverbs

It is quite complex to describe the position of adjectives in a sentence, and for the sake of simplicity, this lesson focuses on the most general rules. Students will have more information and practice in placing adverbs in the right position in *Reward* Intermediate.

Cross-cultural comparison

You will find small differences in the daily routines of the students in your class even if they come from the same macro-culture. These differences can be just as usefully explored as the larger differences between people of different cultures. Remember that culture can involve gender, age, socio-professional status, and religion as well as geographical origin and background.

READING

1 Aim: to prepare for reading.

● Ask the students if they have or watch breakfast television in their countries. Find out how many students watch breakfast television. Ask them what sort of programmes there are.

● Tell the students that Tanya Philips is a breakfast television presenter. Ask them to predict what time Tanya Philips's day begins and ends.

2 Aim: to practise reading for main ideas.

● Ask the students to read the passage and find out if they guessed correctly in 1.

● Try to discourage them from asking about too many vocabulary items. If you explain every word they don't understand, you may increase their vocabulary but you won't develop their ability to deal with reading passages.

Answers

She gets up at half past three and goes to bed at half past eight in the evening.

3 Aim: to practise understanding text organisation.

● This activity type is used several times in the *Reward* series. It is designed to help the students focus on the logical order in which information occurs in a passage, and it is also a motivating activity type. Even if the students don't enjoy the text, it is to be hoped that they will enjoy the task.

Answers

Sentence a goes after ... *and finishes at nine o'clock.*
Sentence b goes after ... *I leave the studio at a quarter past ten.*
Sentence c goes after ... *and I cook dinner.*

4 Aim: to practise reading for specific information; to practise writing questions.

● This activity will encourage the students to think carefully about what was actually said during the interview with Tanya Philips.

● Ask the students to write the questions the journalist asked.

Possible answers

What time does your working day start?
What time do you leave home?
What time do you get to the studio?
What time does the programme start and finish?
What do you do after the programme?
What time do you get home?
What do you do in the afternoon?
What time does your husband get home?
What do you do in the evening?
When do you have dinner?
When do you go to bed?
What do you do at the weekend?

● You may like to ask some students to write the questions for the first half of the passage, and some for the second half.

● Students from each half should ask and answer the questions.

GRAMMAR

1 Aim: to focus on negatives and *-s* endings.

● Ask the students to read the information in the grammar box and then to do the exercises.

● Ask the students to count the number of negatives and *-s* endings.

> **Answers**
> **Negatives:** 2
> ***-s* endings:** 6

2 Aim: to focus on the position of adverbs.

● Ask the students to put the adverbs in the right position. You may like to do this orally with the whole class, but it may be worthwhile for students to write the sentences in full afterwards.

> **Answers**
> 1 I usually get up at seven o'clock.
> 2 I never do the ironing.
> 3 She often has a drink with friends.
> 4 He sometimes goes to bed at eleven o'clock.
> 5 I'm usually asleep at midnight.
> 6 I'm never up at five o'clock in the morning.

3 Aim: to focus on prepositions of time.

● Ask the students to do this activity on their own and then to check it in pairs.

> **Answers**
> 1 from, to 2 at 3 at 4 at 5 on 6 on

VOCABULARY AND SPEAKING

1 Aim: to present the words and expressions in the vocabulary box.

● Ask the students to look at the activities in the box. Make sure everyone understands what the words and expressions mean.

● Ask the students to underline the activities they do every day.

2 Aim: to practise using the expressions in the vocabulary box; to practise speaking.

● Ask the students to work in pairs and to ask and say which activities they do every day.

● Write the activities on the board and count the number of students who do the activities every day.

LISTENING AND WRITING

1 Aim: to prepare for listening.

● Ask the students to look at the list of activities and to say which ones they do every day.

● As they have just done a similar activity with the expressions in the vocabulary box, this activity should be completed quickly.

2 Aim: to listen for main ideas.

● Ask the students to listen to Sam and to tick the activities he does in a typical day.

3 Aim: to check comprehension; to practise using adverbs of frequency.

● Ask the students to work in pairs and to check their answers.

	Sam
go to a club	✓ always
go to a party	✓ always
do some work	
have breakfast	✓ usually
have lunch	✓ usually
go to bed	✓
have dinner in a restaurant	✓ usually
go to a concert	✓ sometimes
meet friends	✓ always
play football	✓ often
telephone a friend	✓ always on Sundays
go to work/school	

4 Aim: to practise writing; to practise using *then, after that, after* + time of day.

● Ask the students to write a magazine article about a typical day in their lives. Make sure they follow the guided writing instructions.

● You may like to ask the students to do this activity for homework.

1 Look at the verbs in the article again. How many are in the negative? How many have *-s* endings?

2 Put the adverbs in brackets in the correct position.

1 I get up at seven o'clock. (usually)
2 I do the ironing. (never)
3 She has a drink with friends. (often)
4 He goes to bed at eleven o'clock. (sometimes)
5 I'm asleep at midnight. (usually)
6 I'm up at five o'clock in the morning. (never)

3 Complete these sentences with a preposition of time.

1 We work _____ Monday _____ Friday.
2 The film starts _____ six o'clock.
3 He leaves home _____ eight o'clock.
4 I do the housework _____ the weekend.
5 She does the ironing _____ Monday.
6 We usually go out _____ Friday evening.

VOCABULARY AND SPEAKING

1 Look at the phrases in the box. Underline the things you always do every day.

go shopping see friends do the washing up
do the housework cook dinner
get the bus/train to work/school/home
do some work/homework go to the cinema

2 Work in pairs. Ask and say what you do every day.

Do you always go shopping? No, I don't.
Do you always do the washing up? Yes, I do.

LISTENING AND WRITING

1 You're going to hear Sam, who lives in London, talking about a typical day in his life. Look at the things he does. Which ones do you always, usually or sometimes do? At what time do you do them? Which ones do you never do?

Sam

go to a club ☐
go to a party ☐
do some work ☐
play music ☐
have breakfast ☐
have lunch ☐
go to bed ☐
have dinner in a restaurant ☐
go to a concert ☐
meet friends ☐
play football ☐
telephone a friend ☐
go to work/school ☐

2 Listen and tick (✓) the things Sam does.

3 Work in pairs. What does Sam always, usually and sometimes do? Can you remember at what time he does these things?

4 Work in pairs. You're going to write a magazine article about a typical day in your life. Say:

– what time you get up
I get up at eight in the morning.
– what you do then
Then I usually have some breakfast.
– what you do after that
After that, I usually read the paper.

Use *then* and *after that* to describe when you do things. Remember to join two phrases with *and*.

12 *How do you get to work?*

Articles (2): *a/an, the* and no article; talking about travel

The age of the train

Which do you prefer, the plane or the train?

We talk to Katie Francis. She's twenty-eight years old and she's a marketing consultant. She lives in London but she often works in France.

How does she get there?

'I go by train. I take the train from Waterloo station through the Channel Tunnel.'

How long does the train take?

'It takes three hours and I work during the journey. By plane, it takes an hour but you have to get to and from Charles de Gaulle airport.'

What time does she leave London?

'There are several trains a day. I get the train at eight o'clock and arrive at the Gare du Nord station in Paris at midday French time. I work in an office which is very close, so I go there on foot. I'm at work by half past twelve.'

And when does she leave for home?

'I usually stay at a friend's flat, work next day and get the seven o'clock train home. I get to London at nine o'clock British time, and I'm home by a quarter to ten.'

So what's her opinion of the Channel Tunnel service?

'It's great. It's definitely the age of the train.'

VOCABULARY AND READING

1 Work in pairs. Look at the words in the box. Which means of transport can you see in the photo on the opposite page.

train plane tram bicycle underground ferry car

Now turn to Communication activity 17 on page 103 and check you know what the other words mean.

2 Match the means of transport in 1 with the following places.

station airport garage bus stop

3 Which verbs go with the different means of transport?

drive ride take

4 Which is the odd word out?

1 train bus stop station airport
2 ride bike tram
3 car bicycle tram drive
4 underground ferry train airport

5 Match the adjectives with their opposites. Which adjective has no opposite in the box?

cheap expensive far fast near slow crowded

cheap – expensive

6 Look at the photo of Katie Francis and the title of the article. What do you think *The age of the train* means?

1 Trains are very slow.
2 Trains are very old.
3 Trains are very popular today.

Now read the article and find out if you guessed correctly.

12

GENERAL COMMENTS

The article

For speakers of many languages, especially for those whose mother tongue has no article system, the article system in English is very complex. This lesson focuses on a few of the most useful rules about the definite, indefinite and the so-called zero article. Many teachers will assume that it's better to cover the use of articles on an *ad hoc* basis, relying on exposure and acquisition, rather than formal teaching. There is much in favour of this approach, but it is felt that a simple lesson such as this is useful in drawing attention to the issue and raising some of the difficulties.

VOCABULARY AND READING

1 Aim: to present the words in the vocabulary box.

- Ask the students to look at the photo and to say what kind of transport they can see. This may be a good opportunity to reactivate passive knowledge of the words to describe kinds of transport.

Answer
train

- You may like to explain that the train is the Eurostar, the train which connects London to Paris and Brussels via the Channel Tunnel. It is a *train à grande vitesse (TGV)*, which reaches speeds of over 250km/h.

- Explain that the Communication activity will contain photos showing the other kinds of transport.

2 Aim: to present the words in the vocabulary box.

- The words in the box are the places which are linked with the kinds of transport in 1.

- Ask the students to match the places and the kinds of transport.

Answer
station: train, underground
airport: plane
garage: car
bus stop: tram

3 Aim: to present the words in the vocabulary box.

- This is the last of the three activities focusing on collocations in the word field of transport.

- Ask the students to match the verbs with the means of transport.

Answers
drive: a car, a tram, a bus
ride: a bicycle
take: a tram, a train, a plane, a ferry, the underground

4 Aim: to check comprehension.

- Explain that there is one word in each group of words which doesn't belong. Ask the students to decide which is the odd word out.

Answers
1 train 2 ride 3 drive 4 airport

- Can the students explain why these words are the odd ones out?

5 Aim: to present the words in the vocabulary box.

- Write these adjectives on the board.
  cheap far fast

- Ask the students to match the adjectives and their opposites.

Answers
cheap - expensive, far - near, fast - slow
The word without an opposite is *crowded.*

6 Aim: to prepare for reading; to practise reading for main ideas.

- Encourage the students to think about the title of the passage and to predict what it might be about.

Answer
3

- Ask the students to quickly read the passage and see if they guessed correctly.

7 Aim: to practise reading for specific information.

- Ask the students to work in pairs and to answer the questions.

- If there's time, do this activity in writing.

Answers
1 She's a marketing consultant.
2 She lives in London.
3 She takes the train.
4 It takes three hours.
5 She leaves London at eight o'clock.
6 She arrives in Paris at midday, French time.
7 She stays at a friend's flat.
8 Yes, she does.

8 Aim: to practise inferring.

- This activity encourages the students to read between the lines and to infer the advantages from what Katie says.

Answers
Advantages:
It only takes three hours.
She works during the journey.
Her office is close to the station in Paris.
She can walk to her office.

GRAMMAR AND FUNCTIONS

1 Aim: to focus on the uses of the article.

- Ask the students to read the information in the grammar box and then to do the exercises.
- You may need to give the students a lot of help with this activity. It takes an inductive approach to the presentation of the uses of the article.

Answers
Indefinite article with jobs.
She's a marketing consultant.
Indefinite article when you talk about something for the first time.
I work in an office.
Definite article when you talk about something again.
I work during the journey.
Definite article when there is only one.
I take the Eurostar service.
No article with most countries.
She works in France.
No article with certain expressions.
I *go by train. I go there on foot.*

2 Aim: to practise using the article.

- You may like to do this activity orally.

Answers
Georges Bastien is **a** ticket inspector. He works on **the** Eurostar service between Britain, France and Belgium. He travels to London and Brussels several times **a** week. He lives in **a** house in Paris. **The** house is about thirty minutes from **the** Gare du Nord. And
how does he get to work? **By** bicycle!

3 Aim: to practise talking about travel.

- Ask the students to do this activity in pairs and match the questions and answers.
- Ask the students to work in pairs and ask and answer the questions.

Answers
1e 2d 3a 4b 5c

SOUNDS

1 Aim: to focus on the pronunciation of *the* before vowels and consonants.

- This is another inductive exercise to encourage the students to formulate the rules for themselves.
- Say some of the words aloud and make sure you say /ðə/ before a consonant and /ði:/ before a vowel.
- Play the tape and ask the students to listen.
- Ask the students to formulate a rule for the correct pronunciation.

Answer
You say /ðə/ before a consonant, and /ði:/ before a vowel.

2 Aim: to practise the correct pronunciation of *the*.

- Ask the students to put the words in two columns. You can write /ðə/ and /ði:/ on the board and ask the students to come up and write the words under the correct phonetic transcription.
- Play the tape for the students to check their answers.

Answer
/ðə/: the teacher, the book, the video
/ði:/: the answer, the uncle, the aunt, the evening, the afternoon

SPEAKING AND WRITING

1 Aim: to practise speaking.

- Ask the students to find out how the other students get to school or work every day. They should use the questions in *Grammar and functions* activity 3.
- Make sure they ask several students and note down their answers.

2 Aim: to practise writing.

- Ask the students to write a paragraph describing how one or two of the students get to school or work.
- You may like to ask students to do this for homework.
- Put some of the paragraphs on the wall. Make sure you include a description of every student's journey.

7 Answer the questions.

1 What does Katie Francis do?
2 Where does she live?
3 How does she get to Paris?
4 How long does it take?
5 What time does she leave London?
6 What time does she arrive in Paris?
7 Where does she stay in Paris?
8 Does she like the Eurostar service?

8 What are the advantages of travelling by train for Katie?

It only takes three hours, ...

GRAMMAR AND FUNCTIONS

Articles (2): *a/an*, *the* and no article

You use the indefinite article *a/an*:
– to talk about something for the first time.
*She works in **an** office in Paris.*
– with jobs.
*She's **a** marketing consultant.*
– with certain expressions of quantity.
*There are several trains **a day**.*
You use the definite article *the*:
– to talk about something again.
***The** office is near the Gare du Nord.*
– when there is only one.
***The** Channel Tunnel.*
You don't use an article:
– with most countries.
She goes to France. She lives in Britain.
– with certain expressions.
by train by plane at work at home
(For more information about articles, see Grammar review, page 111.)

Talking about travel
How do you get to work/school?
How long does it take?
How much is it? How far is it?

1 Look for examples of the indefinite and definite article in *The age of the train*. Which rules in the grammar box do they show?

2 Complete the sentences with *a/an* or *the*, or put a – if no article is needed.

Georges Bastien is _____ ticket inspector. He works on _____ Eurostar service between _____ Britain, _____ France and _____ Belgium. He travels to London and Brussels several times _____ week. He lives in _____ house in Paris. _____ house is about thirty minutes from _____ Gare du Nord. And how does he get to _____ work? _____ bicycle!

3 Match the questions and the answers.

1 How do you get to work/school?
2 How long does it take?
3 How much is it?
4 How far is it?
5 Is it crowded?

a Three pounds. It's quite expensive.
b It's about three kilometres.
c No, it isn't.
d It takes about thirty minutes.
e I go by bus.

SOUNDS

1 Listen to these words. Notice how you pronounce the word *the*.

/ðə/	/ði:/
the train	the airport
the service	the hour
the station	the office
the bus	the underground

When do you pronounce *the* /ðə/ and when do you pronounce it /ði:/?

Now say the words aloud.

2 Put these words in two columns /ðə/ or /ði:/.

the teacher the answer the book the uncle the aunt the evening the afternoon the video

Listen and check. As you listen, say the words aloud.

SPEAKING AND WRITING

1 Use the questions in *Grammar and functions* activity 3 to find out how other students in your class get to work or school. Make notes.

Fabbio: by bus, ten minutes ...

2 Choose one or two students and write a paragraph using the notes in 1 describing how they get to work or school.

Fabbio goes to school by bus. The journey takes about ten minutes...

13 Can you swim?

Can and _can't_; questions and short answers

VOCABULARY AND READING

1 Match the verbs in the box with the nouns below.

cook do play ride speak use write

a crossword a computer a foreign language
a horse a meal a musical instrument poetry

cook a meal ...

2 Here are some verbs which do not always need nouns after them. Match them with the photos.

draw ski swim type

3 Read *Are you an all-rounder?* and answer the questions with *yes* or *no*.

Are you an all-rounder?

Can you ...

1 run 100 metres in 15 seconds? ☐
2 use a computer? ☐
3 cook a meal for six people? ☐
4 swim 500 metres? ☐
5 speak a foreign language? ☐
6 play a musical instrument? ☐
7 ski? ☐
8 draw? ☐
9 ride a horse? ☐
10 write poetry? ☐
11 drive a car? ☐
12 knit a jumper? ☐
13 type quickly? ☐
14 do crosswords? ☐

4 Count your *Yes* answers. Now turn to Communication activity 26 on page 105 to find out if you're an all-rounder.

GRAMMAR

Can
***Can* is a modal verb.**
You use *can:*
– to talk about something you are able to do on most occasions.
*I **can** drive.*

Can't
***Can't* is the negative of *can*.**
*I **can't** ride a bike.*
(For more information about modal verbs, see Grammar review page 111.)

Questions and short answers
Can you speak English? *Yes, I can.* *No, I can't.*
You use *So can I* and *Nor can I* to describe the same abilities as someone else.
I can drive. So can I. *I can't cook. Nor can I.*

13

GENERAL COMMENTS

Types of listening material

Most of the listening material in *Reward* Elementary is scripted and carefully graded to the level of the student. In the *Speaking and listening* section of this lesson, however, there is a type of listening material which is used extensively in later books of the *Reward* series, and which is referred to as semi-scripted material. The passage will contain a certain amount of extraneous language which may or may not be above the students' present level, but which has an authentic sound to it, as if the students were listening to real-life, everyday English. This kind of listening material is recorded in a studio, to ensure good acoustic quality, and involves actors responding to a prompt of some kind. In this case, the prompt is the statements you see in *Speaking and listening* activity 1.

If the students are only exposed to scripted material, they will not be prepared for real-life conditions. If they are exposed to authentic material, they will suffer a serious loss of confidence, and will only register how little, rather than how much, they understand. Furthermore, the quality of the recording may not be very good. Semi-scripted material is an attempt to retain the advantages of scripted and authentic material, and avoid their disadvantages.

VOCABULARY AND READING

1 Aim: to present the words in the vocabulary box.

- There has already been some collocation work in earlier lessons. This activity focuses once again on words which go together. Write the verbs in the box on the board, and say them aloud. Ask the students to suggest words and expressions which go with the verbs.

- Ask the students to suggest other words and expressions which go with the verbs.

Answers
cook a meal, do a crossword, play a musical instrument, ride a horse, speak a foreign language, use a computer, write poetry

2 Aim: to present the words in the vocabulary box.

- These verbs can be used intransitively, that is, without an object. Ask the students to match them with the photos.

Answers
a type
b swim
c ski
d draw

3 Aim: to practise using the new vocabulary; to practise reading and reacting to a text.

- You may like to explain that an *all-rounder* is someone who is good at many different things.

- Ask the students to read the questionnaire and to put either *yes* or *no* by the things which they can do.

4 Aim: to continue the reading practice.

- The fact that the questionnaire is in the second person and that there is a scoring system should ensure that the students' motivation is engaged.

- Ask the students to add up their *yes* answers and to turn to the Communication activity as advised. Of course, they shouldn't take the analysis of their score too seriously!

GRAMMAR

1 Aim: to practise short questions and answers with *can*.

- Ask the students to read the information in the grammar box and then to do the activities.

- Ask the students to do this activity orally in pairs.

- Check the answers with the whole class.

Answers
1 Can you swim? Yes, I can.
2 I can drive, but I can't cook.
3 I can't ski. Nor can I.
4 I can draw. So can I.
5 Can you do crosswords? No, I can't.
6 I can't drive. Nor can I.

2 Aim: to practise using the target structures.

- Ask the students to do this activity on their own.
- Ask the students to work in pairs and to check their answers by acting out the conversation.
- Play the tape and ask the students to check their answers.

Answers
1d 2c 3a 4b

3 Aim: to practise speaking.

- Ask the students to work in new pairs and to act out the conversation.
- Ask one or two pairs to act out the conversation to the whole class.

4 Aim: to practise speaking.

- Ask the students to talk about their answers to the questionnaire. The conversation in activity 2 can be used as a model.
- Ask one or two pairs to act out the conversation to the whole class.

SOUNDS

1 Aim: to focus on the unstressed pronunciation of *can* and *can't*.

- When *can* is isolated at the end of a sentence and to a lesser extent, at the beginning of a sentence, it is pronounced /kæn/. When it's followed by an infinitive, its unstressed pronunciation is /kən/.
- Play the tape and ask the students to decide which they hear. Of course the correct answer to this activity can be predicted by reading the transcript beforehand.

Answers

1 /kæn/, /kæn/	4 /kən/, /kən/
2 /kən/, /kɑːnt/	5 /kæn/, /kɑːnt/
3 /kɑːnt/, /kən/	6 /kɑːnt/, /kən/

Aim: to focus on the unstressed pronunciation of *can* and *can't*.

- Ask the students to say the sentences aloud.
- You can continue this work by asking students to predict the pronunciation of *can* and *can't* in *Grammar* activity 2, and then play the tape to check.

SPEAKING AND LISTENING

1 Aim: to practise reading for main ideas; to check comprehension; to prepare for speaking and listening.

- These statements constitute a minimal amount of reading but they need to be fully understood if they are to act as a listening comprehension check and as a stimulus for speaking.
- Ask your students to read them and match them with the drawings. There are of course several which aren't illustrated.

2 Aim: to practise speaking.

- Ask the students to work in pairs and to discuss which ones are true.

3 Aim: to practise listening for main ideas.

- Ask the students to listen to two people discussing whether the statements are true or false. Explain that the speakers will use natural, everyday English, so there may be words and expressions which the students don't understand. Stress, however, that it will be clear and comprehensible when the speakers say if the statements are true or false.
- Play the tape.

4 Aim: to practise speaking and listening.

- Ask the students to work in pairs and to check their answers.

Answers

T You can see the Great Wall of China from space.
F Cats can't swim.
T Chickens can't fly.
F Computers can write novels.
F Cameras can't lie.
T England can win the nextWorld Cup.
F Thin people can't swim very well.
F You can never read a doctor's handwriting.
T You can clean coins with Coca Cola.
T Cats can see in the dark.

- Play the tape again.

5 Aim: to practise speaking.

- Ask the students to work in pairs. Encourage them to write statements about people in the class or people which everyone knows. They should write two true statements and one false one.

6 Aim: to practise speaking.

- Students should now exchange their statements and read the ones they have received from another pair. They should try to guess which statement is false.

1 Complete the sentences.

1 _____ you swim? Yes, I _____
2 I _____ drive, but I can't cook.
3 I can't ski. _____ can I.
4 I can draw. _____ can I.
5 Can you do crosswords?
No, _____
6 I can't drive. _____ can I.

2 Decide where the sentences a – d go in this conversation.

JO What can you do?
PAT (1) _____.
JO So can I.
PAT (2) _____
JO Nor can I. But I can speak a foreign language.
PAT (3) _____
JO Spanish. Can you speak Spanish?
PAT (4) _____

a What language can you speak?
b No, I can't.
c But I can't cook very well.
d I can run a 100 metres in 15 seconds and I can use a computer.

Now listen and check.

3 Work in pairs and act out the conversation.

4 Work in pairs. Ask and say what things in the questionnaire you can do. Use the conversation in 2 to help you.

SOUNDS

1 Listen to the sentences in *Grammar* activity 1. Do you hear /cən/, /cæn/ or /cɑ:nt/?

2 Now say the sentences aloud.

SPEAKING AND LISTENING

1 Look at these statements. Which statements can you see in the pictures?

☐ You can see the Great Wall of China from space.
☐ Cats can't swim.
☐ Chickens can't fly.
☐ Computers can write novels.
☐ Cameras can't lie.
☐ England can win the next World Cup competition.
☐ Thin people can't swim very well.
☐ You can never read a doctor's handwriting.
☐ You can clean coins with Coca Cola.
☐ Cats can see in the dark.

2 Work in pairs. Which statements are true?

3 Listen to Ann and Frank discussing the statements. Tick (✓) the ones they think are true.

4 Work in pairs and check your answers to 3.

Now listen again and check.

5 Work in pairs. Write two true statements and one false statement. Use *can/can't*.

6 Work with another pair. Give your statements to them and look at their statements. Try and guess which statement is false.

Max can ski.
Jenny can't drive.
Paco can run 100 metres in nine seconds.

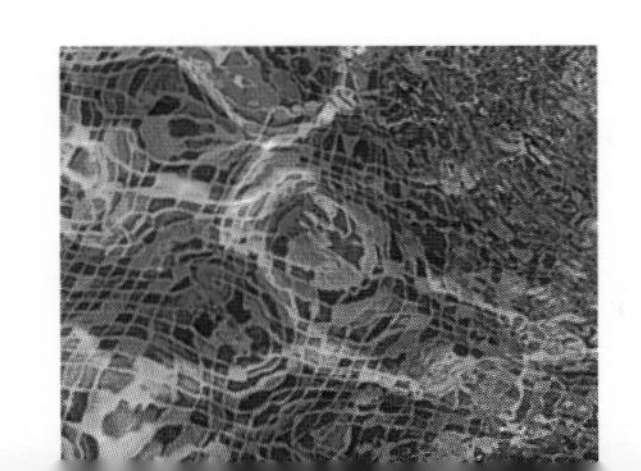

14 How do I get to Queen Street?

Prepositions of place (3); asking for and giving directions

VOCABULARY AND LISTENING

1 Complete the sentences below with words from the box.

baker bank bookshop car park chemist cinema florist greengrocer library market newsagent phone box post office railway station restaurant supermarket pub swimming pool station taxi rank

1 You can buy stamps in a ____.
2 You can park your car in a ____.
3 You can borrow a book from a ____.
4 You can make a phone call from a ____.
5 You can get a taxi from a ____.
6 You can take the train from a ____.
7 You can go swimming in a ____.
8 You can buy bread at a ____.

1 You can buy stamps in a post office.

2 Work in pairs. Look at the vocabulary box and ask and say where you can:

1 buy some flowers
2 take out some money
3 have a meal
4 buy some aspirin
5 get some vegetables
6 have a drink
7 see a film
8 buy a book

1 Where can you buy some flowers?
At a florist.

3 Look at the map and listen to four conversations. Mark the places on the map.

GRAMMAR AND FUNCTIONS

Prepositions of place (3)

There's a florist ***next to*** *the bank.*
There's a bank ***between*** *the florist and the post office.*
There's a bookshop ***on*** *the corner of Queen Street.*
There's a chemist ***opposite*** *the restaurant.*

Asking for and giving directions

Excuse me, where's the station? How do I get to Queen Street?
Go along Prince Street. Across East Street. Up George Street. Down Valley Road.
Turn left into George Street. Turn right into Queen Street.
It's on the left. It's on the right. It's straight ahead.

1 Work in pairs. Check your answers to *Vocabulary and listening* activity 3.

There's a cinema opposite the bank.

2 Look at the map and say where other places are.

There's a taxi rank opposite the swimming pool.

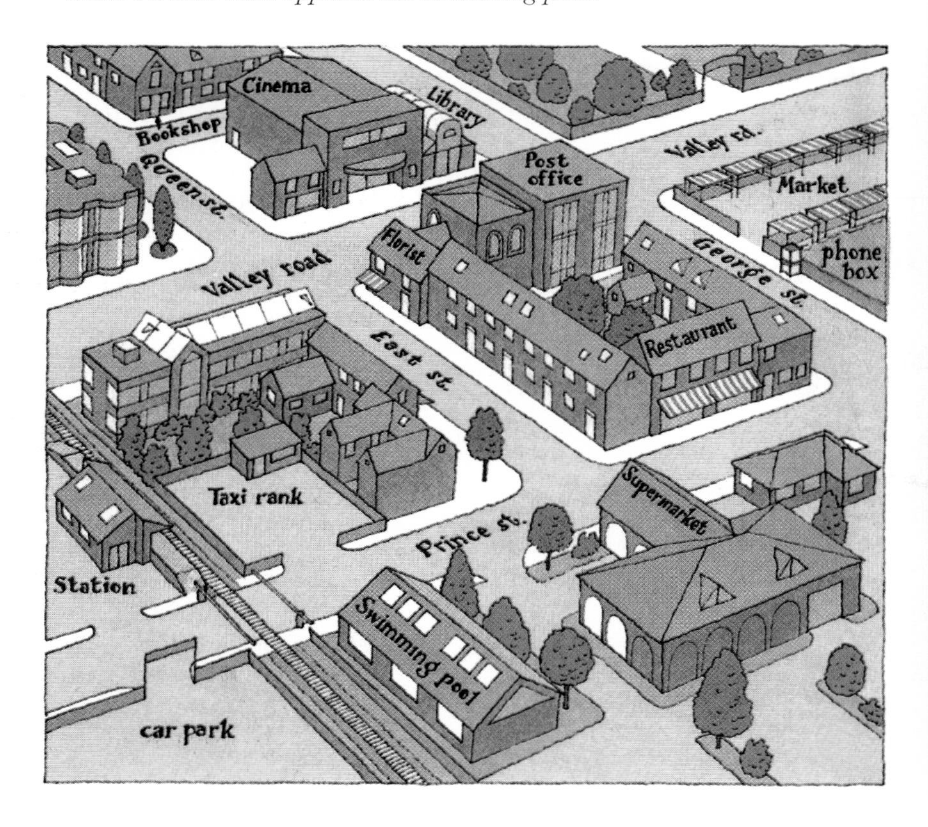

14

GENERAL COMMENTS

Prepositions

It may be useful at this stage to spend a few minutes at the beginning of the lesson revising the prepositions the students have come across already: *to, at, in, from, under, near*. Make sure students understand that *at* and *in* are also prepositions of time as well as place.

VOCABULARY AND LISTENING

1 Aim: to present the words in the vocabulary box.

- Ask the students to give examples of shops and town facilities near their school using the vocabulary in the box. If you're teaching in a school near a shopping centre or high street, you may like to draw a map of the area on the board and label the shops you can find there.

- Ask the students to complete the sentences with words from the box.

Answers
1 post office 2 car park 3 library
4 phone box 5 taxi rank 6 railway station
7 swimming pool 8 baker's

2 Aim: to practise using the new vocabulary.

- Ask the students to work in pairs and to say where you can buy the things shown.

- Check the answers with the whole class.

Answers
1 florist 2 bank 3 restaurant
4 chemist 5 greengrocer 6 pub
7 cinema 8 bookshop

3 Aim: to practise listening for specific information.

- Explain to the students that they are going to hear four conversations in the street and that they need to mark the names of the places mentioned on the map.

- Play the tape.

Answers
The four places are: a bank, a baker's, a chemist, a newsagent's

GRAMMAR AND FUNCTIONS

1 Aim: to check listening comprehension; to practise using prepositions of place.

- Ask the students to read the information in the grammar and functions box and then to do the exercises.

- You may need to illustrate *next to, between, on the corner of* and *opposite* with drawings on the board. If you haven't drawn a map of your own local shopping district, it may be a good idea if you quickly copy the map in the Student's Book on the board, so you can maintain visual contact with your students.

Answers
1 The cinema is opposite the bank.
2 The baker's is opposite the florist.
3 The chemist's is opposite the restaurant.
4 The newsagent's is in East Street.

2 Aim: to practise using prepositions of place.

- Ask the students to continue to say where the shops and other facilities are on the map.

- You can continue this activity by referring back to the map of your local shopping district.

3 Aim: to present the language for giving directions; to practise understanding text organisation; to provide a model conversation.

- Ask the students to read the sentences from the conversation and to put them in the right order.
- They can check their answers in pairs.
- Play the tape.

Answers
d e b c a

4 Aim: to practise giving directions.

- Ask the students to act out the conversation in pairs.
- When they've done this a few times with different partners, ask two or three pairs to act it out in front of the whole class.

5 Aim: to practise using prepositions of place and giving instructions.

- Ask the students to work in pairs and to follow the instructions in the Communication activities.
- Tell the students that each pair has the same map, but has different information about the shops and town facilities marked on the map.

READING AND SPEAKING

1 Aim: to prepare for reading; to pre-teach some new words.

- This activity is designed to start the students thinking about the passage they are going to read.
- You may want to explain that the *check-out* is where you pay for goods in a supermarket or shop, and that a *shopping mall* is an American English expression for a commercial centre, with many individual shops in one huge building.

2 Aim: to practise reading for main ideas.

- Explain to the students that the passage has one central idea, and that it's important they should understand what it is. Explain that *The end of the High Street?* questions whether the High Street as a shopping centre will survive for much longer.
- Ask the students to read the passage and to decide why it is the end of the High Street.

Answer
1

- Go back over the passage and check everyone understands why this is the correct answer.

3 Aim: to practise reading for specific information.

- These questions are designed to make the students go back over the passage and read it more closely.

Answers
1 True 2 False 3 True 4 True 5 False 6 True

- Students may like to check their answers in pairs.

4 Aim: to practise speaking; to check comprehension.

- Ask the students to discuss their answers to the questions in relation to their own country.

3 Here's a conversation from *Vocabulary and listening* activity 3. Put the sentences in the right order.

a Thank you.
b How do I get to East Street?
c Go along Prince Street. Turn right into East Street. The newsagent is on your left.
d Excuse me, where can I buy a newspaper?
e There's a newsagent in East Street.

Now listen and check.

4 Work in pairs. Act out conversations. Use your answers to *Vocabulary and listening* activity 3 and the conversation above to help you.

Excuse me, where can I buy ...?

5 Work in pairs.

Student A: Turn to Communication activity 4 on page 99.

Student B: Turn to Communication activity 12 on page 101.

The end of the High Street?

In the High Street of a British town, there are shops, banks, cinemas and the post office. There's usually the baker's shop, where you can buy bread, the greengrocer's where you can buy fruit and vegetables and the newsagent's, where you can buy newspapers and magazines. In many towns there is also a food market once or twice a week.

In a High Street shop you usually tell an assistant what you want, then you pay the assistant. In a supermarket you choose what you want, and then take it to the check-out and pay there. There is often a queue of people at the check-out.

There are many kinds of restaurants – you often see Italian, Chinese and Indian restaurants. You don't often see people eating outside because it's too cold.

But now there are shopping malls everywhere. They are usually outside the town with all the shops you see in the High Street. You drive there, leave your car in the car park and go shopping. Today shopping malls are very popular. Is this the end of the High Street?

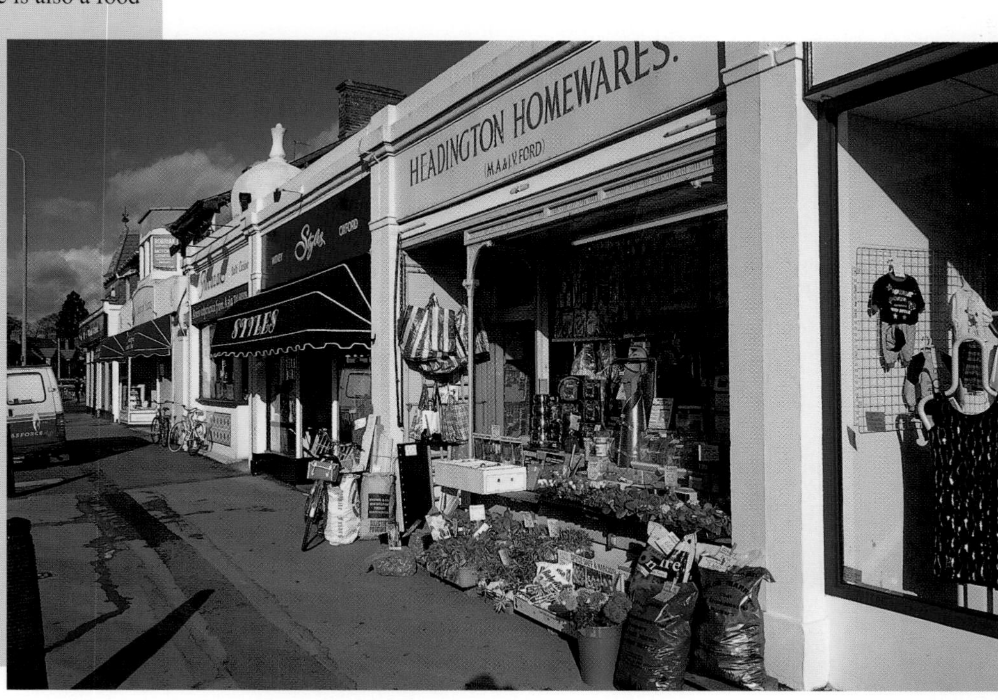

READING AND SPEAKING

1 You're going to read about the High Street, which is often the central street for shopping in a British town. Which of these words do you expect to see?

buy check-out assistant journalist queue shopping mall popular

2 Read the passage. Why is it the end of the High Street? Choose one of the following reasons.

1 Because the shopping malls outside towns are very popular.
2 Because of the food markets once or twice a week.
3 Because there are few assistants in the shops.

3 Read these statements about the passage. Are they true or false for Britain?

1 There are many small shops.
2 There is a market every day.
3 There are assistants to help you.
4 There are often queues in shops.
5 People often eat outside.
6 Many people go shopping in shopping malls.

4 Work in pairs and say if the statements in 3 are true for your country.

15 What's happening?

Present continuous

LISTENING

1 Read the pilot's announcement and decide when it is taking place:
– on take off – during the flight – on landing

'**Ladies and gentlemen, this is your pilot speaking. I hope you're enjoying your flight to Milan this morning. At the moment, we're passing over the beautiful city of Geneva, in the east of Switzerland. If you're sitting on the right-hand side of the plane, you can see the city from the window. We're flying at 10,000 metres and we're travelling at a speed of 700 km/h. I'm afraid the weather in Milan this morning is not very good. It's raining and there's a light wind blowing. Enjoy the rest of your flight. Thank you for flying with us today.**'

2 Listen and underline anything which is different from what you hear.

3 Work in pairs and correct the announcement in 1.

4 Listen again and check your answers to 3.

GRAMMAR

Present continuous

You use the present continuous to say what is happening now or around now.

*We'**re flying** at 10,000 metres. It'**s raining** in Milan.*

You form the present continuous with *am/is/are* + present participle (verb + *-ing*). (For more information see Grammar review, page 107.)

Questions	Short answers	Negatives
Is she learning to swim?	*Yes, she is. No, she isn't.*	*She isn't learning to swim.*
Are you enjoying the flight?	*Yes, I am. No, I'm not.*	*I'm not enjoying the flight.*

1 Look at your corrected version of the announcement in *Listening* activity 1 and complete the sentences.

1 The plane _____ going to Rome.
2 They're _____ over Zurich at the moment.
3 They _____ flying at 12,000 metres, they're _____ at 10,000 metres.
4 They aren't travelling at 700 km/h, they _____ travelling at 750 km/h.
5 It _____ raining in Rome. It _____ snowing.

2 Listen to three conversations and decide what the situation is. Choose from the following.

– in an office – in a restaurant
– in the market – on a bus
– in the street – in a gallery

3 Say what the people are doing in the situations in 2. Use the verbs below.

use a computer have a meal
buy food go to work
do the shopping look at paintings

In the first conversation, they're having a meal.

15

GENERAL COMMENTS

Present continuous

This is the first time the present continuous is introduced in *Reward* Elementary, although students may have come across the tense during another course. Its main use in its present sense is to stress the temporary or ongoing nature of what is being described, as opposed to something which is finished or permanent. Its most common use, however, is reserved for Lesson 24: to describe something decided in the future, especially with *going to*.

LISTENING

1 Aim: to present the use of the present continuous; to prepare for listening.

- The choice of context and material is designed to reflect the principal use of the present continuous: to say what is happening at the moment.

- Ask the students if they have ever been in an aeroplane and if so, if they can remember some of the announcements made during the flight.

- Ask the students to read the passage and decide when it is taking place.

> **Answer**
> during the flight

2 Aim: to practise listening for specific information.

- This activity involves some careful listening. Play the tape and ask the students to underline anything in the passage which is different from what they hear.

> **Answers**
> Ladies and gentlemen, this is your captain speaking. I hope you're enjoying your flight to Rome this morning. At the moment, we're flying over the beautiful city of Zurich, in the centre of Switzerland. If you're sitting on the left-hand side of the plane, you can see the city from the window. We're flying at 12,000 metres and we're flying at a speed of 750 km/h. I'm afraid the weather in Rome this morning is not very good. It's snowing and there's a light wind blowing. Enjoy the rest of your flight. Thank you for travelling with us today.

3 Aim: to check comprehension.

- Ask the students to check their answers in pairs. They should change the reading passage to what they've heard on the tape.

- You may like to ask one or two students to read out their corrected versions.

4 Aim: to provide an opportunity for a second listening.

- Play the tape again and ask the students to check their version.

- As far as you, the teacher, are concerned, it doesn't matter if the students' versions are not completely accurate. The process of listening, underlining differences and checking is more important than the final result. However, the challenge to the students to get a correct version will be a powerful motivating force.

GRAMMAR

1 Aim: to focus on the form of the present continuous.

- Ask the students to read the information in the grammar box and then to do the exercises.

- Ask the students to complete the sentences. You may want to do this orally with the whole class.

> **Answers**
> 1 is, 2 flying 3 are, not flying 4 are, 5 isn't, is

2 Aim: to practise using the present continuous.

- Ask the students to listen and decide what the situation is. Emphasise that they do not have to understand every single word. Even if the language is difficult, there will still be something which is comprehensible to them, and in this case, it should be the situation.

> **Answers**
> 1 in a restaurant
> 2 in the market
> 3 on a bus

3 Aim: to practise using the present continuous.

- Ask the students to say what the people are doing in the situations they heard.

> **Answers**
> 1 They're having a meal.
> 2 They're buying food./They're doing the shopping.
> 3 They're going to work.

SOUNDS

1 Aim: to practise the /ŋ/ sound.

- Play the tape and ask the students to listen and repeat the present participles.
- Ask two or three students to say the participles to the whole class.
- Ask the students if the phoneme exists in their own language.

2 Aim: to focus on word stress in sentences.

- If the students manage to understand this basic aspect of pronunciation in English, there will be a backwash effect on listening comprehension as well.
- Ask the students to predict which words are likely to be stressed.

3 Aim: to focus on word stress in sentences.

- Play the tape and ask the students to check their answers. As they do this, they should say the sentences aloud.

Possible answers
This is your captain speaking.
I hope you're enjoying your flight.
We're passing over the beautiful city of Zurich.
You can see the city from the window.

- It is sometimes difficult to get the exact words which the speaker stresses. This becomes easier with more practice, but the students and even native speakers may never become one hundred per cent accurate at this activity. Drawing the students' attention to this aspect of English on a regular basis will nevertheless be a positive experience.

VOCABULARY AND SPEAKING

1 Aim: to present the words in the vocabulary box.

- Some languages and cultures do not use the compass points as frequently as British and American speakers. Find out if your students come from a culture where they use them to describe, for example, different districts in a city, or different regions of their country.
- Ask the students to label the compass with the words for compass points.

Answers

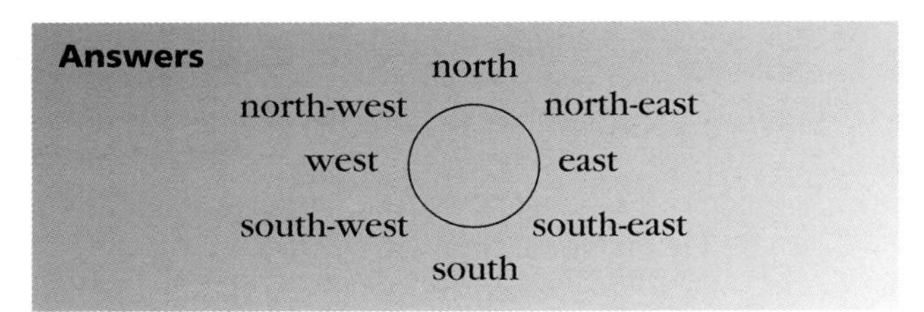

2 Aim: to practise using the new vocabulary.

- Ask the students to describe the location of various places, including the capital city, using the compass points.

3 Aim: to practise using the new vocabulary; to present the words and expressions in the vocabulary box.

- Ask the students to match the words in the box to the places mentioned. If they don't know of these places, choose other places which they do know.
- Ask the students to use the words and expressions to describe the town they're in at the moment.

4 Aim: to practise using the new vocabulary.

- Ask the students to work in pairs and to practise describing various places.

READING AND WRITING

1 Aim: to practise reading for specific information; to prepare for writing; to practise using the present continuous.

- Find out who sends postcards when they're away from home, and to whom. How many are there in the class?
- Ask the students to read the postcard and to answer the questions.

Answers
1 Janet
2 in a villa in the mountains
3 in the hotel in the picture
4 to Fiona Graham
5 yes, they are.

2 Aim: to practise speaking; to check comprehension.

- Ask the students to work in pairs and to check their answers.

3 Aim: to practise writing.

- Ask the students to use the postcard as a model, to follow the instructions and to write a postcard to a friend. They can either invent a holiday location or they can honestly describe where they are and what they're doing at the moment.
- You may like to ask the students to do this activity for homework.

SOUNDS

1 Listen and repeat these present participles.

having watching lying learning
going shopping buying looking
playing reading talking

2 The stressed words in spoken English are the words the speaker thinks are important. Look at these sentences from the pilot's announcement. Underline the stressed words.

1 This is your captain speaking.
2 I hope you're enjoying your flight.
3 We're passing over the beautiful city of Zurich.
4 You can see the city from the window.

3 Listen and check. As you listen, say the sentences aloud.

VOCABULARY AND SPEAKING

1 Use the words in the box to label the compass.

south-west	south-east
north-west	north-east

2 Work in pairs. Say where the capital city is in your country.

London is in the south-east of England.

3 Choose words and phrases in the box to describe four of these places.

Marseille New York Moscow Rio Barcelona Bangkok Tokyo
Venice your town

beautiful attractive on the coast interesting on the river ugly in the mountains boring industrial lively lovely modern old on an island big small

Marseille – lively, on the coast.

4 Work in pairs. Think of other places you can describe using the words in the box.

READING AND WRITING

1 Read the postcard and answer the questions.

– who's writing the postcard?
– where are they staying?
– where's she writing it?
– who's she writing it to?
– are they enjoying themselves?

Dear Fiona,

Here we are on the island of Mykonos in the south Aegean sea, and we're having a wonderful time. We're staying in a villa in the mountains, near a beautiful beach. I'm writing this postcard in the hotel you can see in the picture. Jacqui is learning to swim, and Terry is lying in the sun at the moment. We're enjoying the holiday very much.

See you soon!

Love, Janet and the kids

Fiona Graham
22, Park Street
Stow-on-the-Wold
Gloucestershire
ENGLAND

2 Work in pairs and check your answers to 1.

3 Write a postcard to a friend. Use the postcard in 1 to help you.

Say:
– where you are
– where you're staying
– where you're writing the card
– what you're doing
– if you're enjoying yourself

Progress check 11–15

VOCABULARY

1 There are many international words in English. Here are some types of international words.

food – *burger pizza*
places of entertainment – *theatre cinema*
types of entertainment – *video music*
sports – *football tennis*
jobs – *president doctor*
brand names – *Coca Cola McDonalds*

Are these words also in your language?

2 Put these international words in the types shown in 1.

Pepsi Toyota secretary restaurant pasta handball

3 There are some words in English which come from other languages.

siesta concerto spaghetti café judo ballet samba sauna

Are these words also in your language? What languages do you think they come from?

4 Here are some expressions which you use in everyday situations. Match them with the pictures.

Please Thank you Excuse me Pardon Sorry

5 Listen and check what people say.

GRAMMAR

1 Complete the sentences with *a/an* or *the*, or put a – if no article is needed.

1 I'm _____ doctor. What do you do?
2 She's French but she lives in _____ Britain.
3 I go to work by _____ train.
4 _____ office where I work is near here.
5 _____ Alps are in western Europe.
6 I work at _____ home.

2 *A or an?*

teacher writer American Hungarian chair ID card aunt uncle novel airport

3 Put the adverb in brackets in the correct position.

1 I get up at nine on Saturday and Sunday. (usually)
2 I see friends on Saturday evening. (often)
3 I do the shopping on Saturday morning. (always)
4 I go for a walk on Sunday. (sometimes)
5 I watch the football in the park on Sunday morning. (often)
6 I go to bed early on Sunday night. (usually)

1 I usually get up at nine on Saturday and Sunday.

Progress check 11- 15

GENERAL COMMENTS

You can work through this Progress check in the order shown, or concentrate on areas which have caused difficulty in Lessons 11 - 15.

VOCABULARY

1 Aim: to focus on international words.

- This activity is designed to show the students that they have learned a great deal of vocabulary outside the classroom and without necessarily having realised it.

- Spend some time in drawing the students' attention to the number of words which are the same in English and in their language.

2 Aim: to focus on international words.

- Ask the students to decide what type of words the international words are.

Answers
Pepsi, Toyota - brand names
secretary - job
restaurant - place of entertainment
pasta - food
handball - sport

3 Aim: to focus on loan words in English.

- It may be that there are many English words in the students' language. This activity is to show that there are many words in English borrowed from other languages.

- Ask the students to say if the words are in their language, and where they come from.

Answers
siesta - Spanish, concerto - Italian, spaghetti - Italian, café - French, judo - Japanese, ballet - French, samba - Brazilian Portuguese, sauna - Finnish

4 Aim: to present some everyday expressions.

- It's difficult to fit these expressions into the vocabulary boxes in *Reward*, so this opportunity is being taken to present them to the students. They should be words which the students will have come across beforehand, but which they may not have actively learnt.

Answers
a excuse me b pardon c please
d thank you e sorry

5 Aim: to focus on some everyday expressions.

- Play the tape. The students listen and check their answers.

GRAMMAR

1 Aim: to revise articles.

Answers
1 a 2 no article 3 no article 4 The
5 The 6 no article

2 Aim: to revise *a* or *an* before consonants and vowels.

Answers
a teacher, a writer, an American, a Hungarian, a chair, an ID card, an aunt, an uncle, a novel, an airport

3 Aim: to revise the position of adverbs.

Answers
1 I usually get up at nine on Saturday and Sunday.
2 I often see friends on Saturday evening.
3 I always do the shopping on Saturday morning.
4 I sometimes go for a walk on Sunday.
5 I often watch the football in the park on Sunday morning.
6 I usually go to bed early on Sunday night.

4 Aim: to revise question forms.

Answers
1 What time do you get up at the weekend?
2 What do you do on Saturday evening?
3 When do you do the shopping?
4 What do you do on Sunday?
5 What do you do on Sunday morning?
6 What time do you go to bed on Sunday night?

5 Aim: to revise *can* and *can't*.

Answers
1 He can't play the piano.
2 He can drive.
3 He can't use a computer.
4 He can't type.
5 He can do crosswords.
6 He can swim.

6 Aim: to revise prepositions.

- Ask the students to write true sentences describing where places are.

7 Aim: to revise giving directions.

- Ask the students to give true directions to the places they described in 6.

8 Aim: to practise the present continuous.

- Ask the students to write sentences about people and what they're doing at the moment.

SOUNDS

1 Aim: to focus on the pronunciation of /ð/ and /θ/.

- These phonemes may cause some particular difficulties of speakers of certain languages, such as French and Italian, although not Spanish. Ask the students to listen and repeat the words.

2 Aim: to focus on the pronunciation of /ð/ and /θ/.

- Ask the students to predict what these words will sound like. Play the tape. The students check their answers.

Answers
/ð/: their
/θ/: thirty, theatre, thank

3 Aim: to focus on the pronunciation of /ɑ/, /æ/, and /ʌ/.

- Ask the students to listen and repeat the words as you play the tape.

4 Aim: to focus on the pronunciation of /ɑ/, /æ/, and /ʌ/.

- Ask the students to predict which phoneme they're likely to hear. Play the tape. The students check their answers.

Answers
/ɑ/: afternoon bath fast car market
/æ/: hat taxi bank
/ʌ/: run bus

- The answers above show the British received pronunciation of these words. There will be various regional differences.

5 Aim: to focus on stressed words in questions.

- Ask the students to listen and underline the stressed words.

- Ask the students to say the questions aloud.

Answers
How do you get to work?
How long does it take?
Can you speak English?
Do you always cook dinner?

WRITING AND SPEAKING

1 Aim: to focus on the layout of addresses.

- Do this activity with the whole class. It would be useful for you to draw the envelope on the board and to write up the features.

Answers
title: Mr
first name: John
family name: Smith
number of home: 22
road or street: Bailey Close
town: Romsey
county: Hampshire
post code: So53 1JK
country: England

2 Aim: to practise writing addresses.

- Ask the students to lay out the addresses in the English way. Make sure they use suitable punctuation, as well.

Answers
Mrs Hillary Jones
34, Denver Street
Oxford OX4 1SD

Dr Michael Carey
4, Horseferry Road
London E4 2SF

Mr Kenneth Green
45, Golden Hill
Brighton
Sussex BN1 4FG
England

3 Aim: to practise writing addresses.

- Ask the students what the difference is between their own layout of addresses and the layout in English.

4 Aim: to practise writing addresses.

- Ask the students to collect addresses and to write them down.

4 The sentences in 3 are answers to questions. Write the questions.

What time do you get up on Saturday and Sunday?

5 Write sentences saying what Peter can/can't do.

1 play the piano (no)
2 drive (yes)
3 use a computer (no)
4 type (no)
5 do crosswords (yes)
6 swim (yes)

He can't play the piano.

6 Choose four or five places from the list below which are near where you are now, or where you live.

chemist baker supermarket
cinema department store library
station market bank car park

Write sentences saying where they are.

There's a chemist in rue de Rivoli, next to the bookshop.

7 Write directions to the places you describe in 6.

Go along the Champs Elysées and turn left.

8 Think of five people you know. Write sentences saying what they're doing at the moment.

Brigitte is going to work. Frank is taking the train to London.

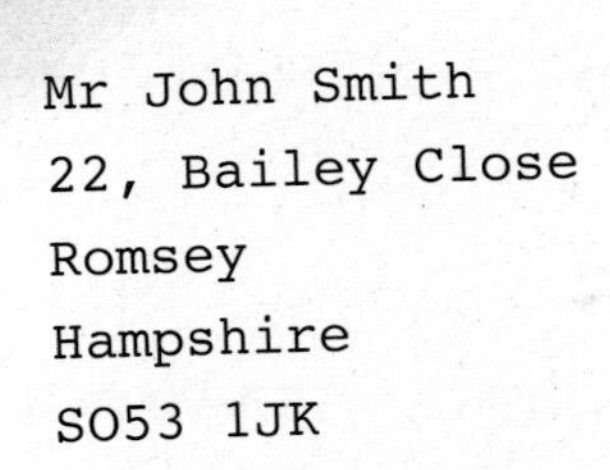

Mr John Smith
22, Bailey Close
Romsey
Hampshire
SO53 1JK
England

SOUNDS

1 Listen to these words. Is the underlined sound /ð/ or /θ/? As you listen, say the words aloud.

this think that three the mother

2 Listen to these words. Is the underlined sound /ð/ or /θ/?

their thirty theatre thank

3 Listen to these words. As you listen, say the words aloud.

start	madam	mother
are	family	Monday
dark	map	lunch
park	have	uncomfortable

4 Listen to these words. Is the underlined sound /ɑː/, /æ/ or /ʌ/?

hat taxi afternoon run bath
fast car bus market bank

5 Listen and underline the stressed words in these questions.

1 How do you get to work?
2 How long does it take?
3 Can you speak English?
4 Do you always cook dinner?

Now say the questions aloud. Make sure you stress the correct words.

WRITING AND SPEAKING

1 Look at the envelope above and find the following features.

title first name family name
county number of home
road or street town country
post code

2 Write the following addresses in a similar way to the address on the envelope in 1.

1 mrs hillary jones 34 denver street oxford ox4 1sd
2 dr michael carey 4 horseferry road london e4 2sf
3 mr kenneth green 45 golden hill brighton sussex bn1 4fg england

3 Work in pairs. Do you write addresses on envelopes in a similar way? What information do you put first?

4 Go round the class asking for addresses of other students. Spell any difficult words.

Who was your first friend?

Past simple (1): *be*

SPEAKING AND LISTENING

1 Think about when you were a child. Look at these questions and think about your answers.

- ☐ Who was your first friend?
- ☐ Where was your first school?
- ☐ What was the name of your first teacher?
- ☐ What was on the walls of your first classroom?
- ☐ What was your best birthday?
- ☐ What was the best day of the week?

2 Listen to four people talking about when they were a child. Put the number of the speaker by the question they answer in 1.

3 Work in pairs. Check your answers to 2.

4 Write your answers to these questions. *When you were a child ...*

- what was your favourite toy?
- what was your favourite game?
- what was your favourite food?
- what was your favourite drink?

5 Your teacher will ask some students to read out their answers to 4. As you listen, tick (✓) any words you hear which are the same as the words you wrote. How many students have the same answers?

GRAMMAR

Past simple (1): *be*

Affirmative	Negative
I was	*I wasn't*
You were	*You weren't*
He/she/it was	*He/she/it wasn't*
We were	*We weren't*
They were	*They weren't*

Questions

*Who **was** your first friend?*
*His name **was** John.*
***Were** you a happy child?*
*Yes, I **was**. No, I **wasn't**.*

1 Look at the grammar box. How many forms does the past simple of the verb *be* have?

2 Work in pairs. Ask and answer the questions in *Speaking and listening* activity 1 and 4.

Who was your first friend?
Her name was Sophia.

3 Write your partner's answers to the questions in 1.

George's first friend was Sophia.

SOUNDS

Listen to these sentences. Do you hear /wɒz/ or /wəz/?

1 My best friend was Jack.
2 Were you born in Rome? Yes, I was.
3 I was quite naughty.
4 My favourite drink was Coke.
5 What was your best birthday?
6 Were you happy? Yes, I was.

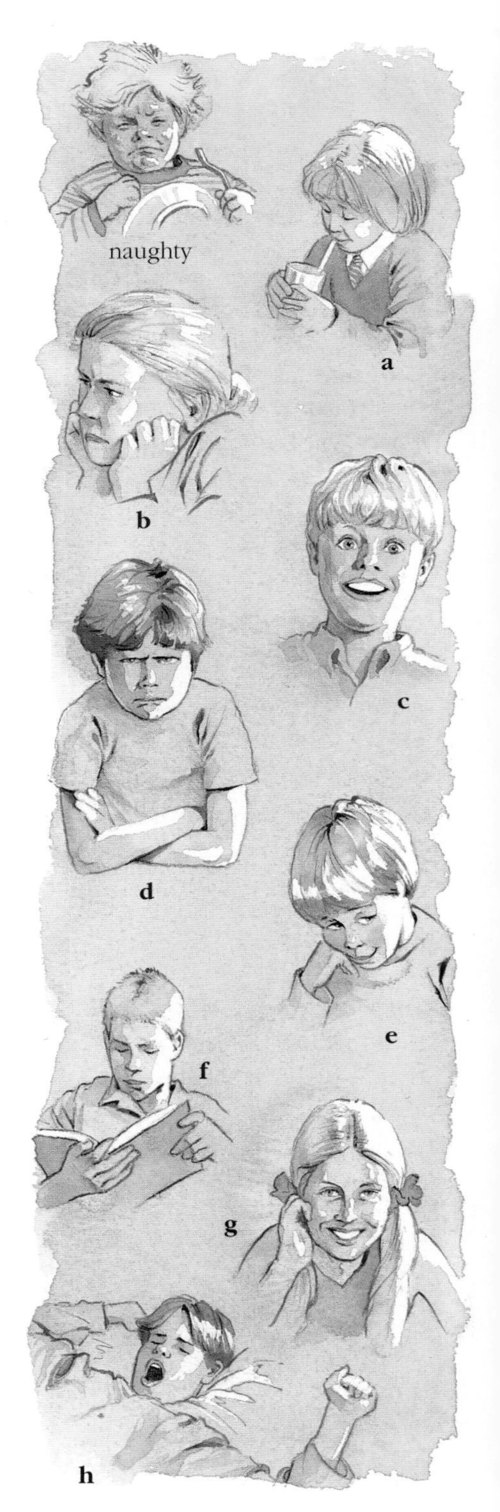

16

GENERAL COMMENTS

Past simple

This is the first of the five lessons in which the past simple is presented. This lesson focuses on the verb *to be*. You may find it useful to restrict the students' discussion of past events to this verb alone in the affirmative and in question form. There will be many opportunities in Lessons 18, 20, 21 and 22 to cover regular and irregular verbs, and negatives.

Respecting the class

The early stages of this activity deals with some highly personal issues and demand the affective involvement of the students. For those classes who are happy to respond, the discussion will be highly motivated. For other classes, you may think it appropriate to leave out some of the questions. The students place is, at present, in the language classroom, not in the psychiatrist's chair!

SPEAKING AND LISTENING

1 Aim: to present the past simple form of *to be*; to prepare for listening.

- Ask the students to think about when they were a child. Ask them to work alone and to think about the questions and their answers to them. It won't yet be suitable to share their answers, because there has not yet been sufficient presentation of the past simple for them to be able to do this.

2 Aim: to practise listening for main ideas.

- The speakers in this listening activity will only talk about some of the questions in activity 1. The students' task is simply to recognise the general idea of their reply and to match it with the question in 1.

- Play the tape.

Answers
Speaker 1: What was the name of your first teacher?
Speaker 2: What was your best birthday?
Speaker 3: What was on the walls on your first classroom?
Speaker 4: Who was your first friend?

- Emphasise that it isn't necessary to understand every word, only the general sense. If you explain every word, you will not help the students to develop their ability to listen to real-life English, which will usually be above their present level.

3 Aim: to check comprehension.

- Ask the students to work in pairs and to check their answers.

- At this stage, you shouldn't expect them to formulate the past simple correctly. This activity focuses on fluency. There will be many opportunities to focus on accuracy later in the lesson.

4 Aim: to prepare for speaking.

- Ask the students to answer the questions in writing. You can go round and help people with any difficult vocabulary. You can also leave out any questions which are not appropriate to the class.

5 Aim: to practise speaking.

- Ask the students to read out their answers to activity 4, if they are willing. See how many people talk about similar experiences and ask the students to tick any words which are similar to the words they wrote.

- Find out how many students have similar answers.

GRAMMAR

1 Aim: to focus on the form of *to be* in the past simple.

- Ask the students to read the information in the grammar box and then to do the exercises.

- This is an inductive exercise to help the students formulate their own rules about grammar.

Answer
Two: ***was*** and **were**

2 Aim: to practise using the past simple of *to be*.

- This should be the first time that the students have to produce the target structure. Do this activity with three or four students in front of the whole class, and then ask them to work in pairs.

- As they do this activity, ask the students to take notes.

3 Aim: to practise using the past simple.

- Ask the students to write their partner's answers in sentences.

- You may like to ask several students to read out their answers to the whole class.

SOUNDS

Aim: to focus on the stressed and unstressed pronunciation of *was*.

- Remind the students that the pronunciation of some words is different according to whether they are stressed or not.

- Play the tape and ask the students to say if they hear /wəz/ or /wɒz/.

Answers
1 /wəz/
2 /wɒz/
3 /wəz/
4 /wəz/
5 /wəz/
6 /wɒz/

VOCABULARY AND SPEAKING

1 Aim: to present the words in the vocabulary box.

● Ask the students to match the words with the drawings.

● Check the students' answers with the whole class. On the board, write the number of the drawing by the word.

Answers
a - well-behaved
b - stubborn
c - happy
d - bad-tempered
e - shy
f - serious
g - friendly
h - lazy

2 Aim: to check comprehension.

● Explain that there is one word in each group of words which doesn't go with the others. Ask the students to underline it. The distinction is always between the positive and negative connotation of the words.

Answers
1 naughty 2 happy 3 well-behaved

● Ask the students if they can explain why these are the odd words out.

3 Aim: to practise using the adjectives in the vocabulary box; to practise using the past simple.

● Ask the students to talk about what they were like as children.

4 Aim: to practise using the adjectives in the vocabulary box; to practise using the past simple.

● Ask the students to identify the people in the photos.

Answers
Queen Elizabeth
Nelson Mandela
Carl Lewis
Madonna

● Then ask the students to say what they think the people in the photos were like as children.

5 Aim: to practise using the adjectives in the vocabulary box; to practise using the past simple.

● Ask the students to write the names of famous people on separate pieces of paper. You may like to have some small squares of paper ready for this activity.

6 Aim: to practise speaking; to practise using the past simple; to practise using the words in the vocabulary box.

● Ask the students to choose one of the pieces of paper and to use the adjectives in the vocabulary box to describe what the person on their piece of paper was like as a child.

7 Aim: to practise speaking; to practise using the past simple; to practise using the words in the vocabulary box.

● Ask the students to read out to the other students in their group the adjectives they have used to describe the person in activity 6. They should not mention the name of the person on their paper. The other students in the group must try and guess the name of the person by the adjectives used to describe him/her.

● Ask several students to read out the name and the list of adjectives they have chosen to the whole class.

WRITING

1 Aim: to practise writing; to practise using the target vocabulary and structures.

● This section is meant to be a light-hearted end to the lesson.

● Ask the students to write statements about themselves as children on a piece of paper. Three statements should be true and one false.

2 Aim: to practise writing; to practise using the target vocabulary and structures.

● Ask the students to show each other the statements and to ask questions about them. They should write the questions on the piece of paper that the statements have been written on.

3 Aim: to practise writing; to practise using the target vocabulary and structures.

● Ask the students to pass back the statements and to read and answer the questions. Encourage them to be as humorous or inventive as they wish.

4 Aim: to practise writing; to practise using the target vocabulary and structures.

● Ask the students to keep asking and answering questions until they have guessed the false statement.

VOCABULARY AND SPEAKING

1 Look at these adjectives to describe character. Match them with the pictures.

naughty happy bad-tempered shy stubborn lazy friendly serious well-behaved

2 Which is the odd word out?

1 happy friendly naughty
2 bad-tempered happy lazy
3 well-behaved lazy stubborn

3 What were you like as a child? Choose adjectives to describe yourself.

Work in pairs. Now ask and say what you were like as a child.

What were you like as a child?
I was quiet, friendly, polite and well-behaved. What were you like?
I was naughty.

4 Work in pairs. Look at the photos of famous people. What do you think they were like as children?

I think the Queen was well-behaved as a child.
Yes, and I think she was serious.

5 Work in groups of four or five. Make a list of six famous people. Write their names on separate pieces of paper and fold them.

6 Choose one of the six pieces of paper. Make a list of adjectives to describe what you think the person on your piece of paper was like as a child.

7 Tell the others what you think the person in 6 was like as a child. Don't say who it is!

I think he was naughty and lazy ...

Can they guess who he/she is?

WRITING

1 Write three true statements and one false statement about you and your family when you were a child.

I was lazy at school.
I was happy at home.
My brother was very naughty.
My teacher's name was Marlon.

2 Work in pairs. Show your partner your statements. Write a question about any of your partner's statements.

My teacher's name was Marlon.
What was your teacher's family name?

3 Write answers to your partner's questions.

His surname was Brando.

4 Continue writing questions about your partner's statements until you guess the false one.

17 How about some oranges?

***Some* and *any* (2); countable and uncountable nouns; making suggestions**

VOCABULARY AND LISTENING

1 Match the words in the box with the photo. Which items can't you see?

apple bacon banana beef bread butter carrot cheese chicken coffee egg grapes lamb lettuce lemon oil onion orange juice potato rice tea tomato water peach cucumber

2 Put the items under these headings:
meat, fruit, vegetables, dairy products, drink.

Which three items are difficult to put under headings?

3 Work in pairs and check your lists in 2. Which items do you often eat or drink?

4 Decide where the sentences a – d go in the conversation.

JEAN OK, what do we need?
TONY We need some fruit and vegetables.
JEAN How about some oranges?
TONY (1) _____
JEAN Yes, there aren't any bananas. And let's get some apples.
TONY OK, apples. And we haven't got any onions.
JEAN (2) _____
TONY That's right, we haven't got any carrots. And let's get some meat.
JEAN Yes, OK. You like chicken, don't you?
TONY (3) _____
JEAN OK two kilos of tomatoes. Anything else?
TONY (4) _____
JEAN No, we need a couple of litres of water and let's get some juice. That's it.

a A kilo of onions. That's enough. And some carrots.
b No. Oh, have we got any water?
c Yes, chicken's great. And we need some tomatoes.
d OK, and we'll have some bananas.

Now listen and check.

GRAMMAR

Some and *any* (2)

You use *some* in affirmative sentences with uncountable nouns and plural nouns.
We need ***some*** *fruit and vegetables.*

You use *some* in questions when you ask for, offer or suggest something.
How about ***some*** *oranges?*

You use *any* in questions and negative sentences with uncountable nouns and plural nouns.
Have we got ***any*** *water? Have we got* ***any*** *oranges?*

Countable and uncountable nouns

Countable nouns have both a singular and a plural form.
a banana – some bananas, a tomato – some tomatoes

Uncountable nouns do not usually have a plural form.
water, juice, coffee

Making suggestions

Let's *have some apples.* ***How about*** *some meat?*

1 Complete the sentences with *some* or *any*.

1 I need _____ water.
2 Have you got _____ onions?
3 I need to do _____ shopping.
4 We need _____ fruit and vegetables.
5 They haven't got _____ potatoes.
6 He wants _____ grapes.
7 How about _____ coffee?
8 There isn't _____ rice.

2 Make a list of everything you ate and drank yesterday. You can use a dictionary if you like.

toast, jam, coffee

Now work in pairs. Ask and say what you ate and drank.

I had some toast and jam, and two cups of coffee.
Did you have any cheese?
Yes, I did.

17

GENERAL COMMENTS

Cultural bias, cross-cultural comparison

As *Reward* is used by teachers around the world, it may be that some of the vocabulary items taught are not the most appropriate for particular cultural contexts. The words for food in this lesson are clearly words which are suitable to describe eating habits in Britain, Europe and North America. You may find it useful to teach these items, but to supplement the vocabulary load with food items appropriate to your own culture. The cultural bias is inevitable but should be seen positively as an opportunity to provide some cross-cultural comparison.

VOCABULARY AND LISTENING

1 Aim: to present the words in the vocabulary box.

- Ask the students to read the list of words and to match them with the items in the photo.
- Check the answers with the whole class.

Answers
coffee, rice, tea, beef, lamb, egg, oil, tomato, bread, butter, cheese, onion, carrot, potato, lettuce, banana, apple, grapes, water, lemon, orange juice

- Ask the students to name the food items which they can't see.

Answers
chicken peach cucumber bacon

2 Aim: to check the meaning of the new vocabulary.

- Ask the students to categorise the food items in the box. You may also like to ask them to do the same with any other new words which you have taught in this lesson.

Answers
meat: bacon, beef, chicken, lamb
fruit: apple, banana, grapes, lemon
vegetables: carrot, lettuce, onion, potato, tomato
dairy produce: butter, cheese, egg
drink: coffee, orange juice, tea, water
The items which are difficult to put under headings are: bread, oil, rice

3 Aim: to practise using the new vocabulary.

- Ask the students which items of food and drink they often eat or drink. This will focus on personal differences of taste and habit, as much as on cross-cultural comparison.

4 Aim: to prepare for listening; to present countable and uncountable nouns.

- Ask the students to read the conversation and to decide where the missing sentences go.

Answers
1d 2a 3c 4b

- Play the tape and ask the students to listen and check.

GRAMMAR

1 Aim: to focus on the use of *some* and *any*.

- Ask the students to read the information in the grammar box and then to do the exercises.
- Ask the students to do this activity in pairs.
- Check the students' answers with the whole class.

Answers
1 some 2 any 3 some 4 some
5 any 6 some 7 some 8 any

2 Aim: to practise using *some* and *any*.

- Ask the students what they ate and drank yesterday. Make a list of the food and drink on the board.
- Ask the students to work in pairs and compare what they ate and drank.

3 Aim: to present countable and uncountable nouns.

● Ask the students if they have countable and uncountable nouns in their own language.

● Ask the students to decide if the nouns in the vocabulary box are countable (C) or uncountable (U).

Answers
Countable: apple, banana, carrot, cucumber, egg, grapes, lettuce, lemon, onion, peach, potato, tomato
Uncountable: bread, bacon, beef, butter, cheese, chicken, coffee, lamb, oil, orange juice, rice, tea, water

4 Aim: to practise using countable and uncountable nouns.

● Ask the students to decide whether the nouns are countable or uncountable. They should then write *some* or *a* before the noun.

Answers
1 a book 2 some money 3 some tea
4 some fruit 5 some meat 6 a pen
7 a sandwich 8 a chair 9 a newspaper
10 some coffee

5 Aim: to practise using the target structures; to practise speaking.

● Ask the students to act out the conversation in *Vocabulary and listening* activity 4.

● Ask two or three pairs to act out the conversation in front of the whole class.

6 Aim: to practise using the target structures; to practise speaking.

● Ask the students to work in pairs and to make a shopping list.

● Ask the students to act out a conversation based on the one in *Vocabulary and listening* activity 4.

SOUNDS

1 Aim: to focus on /s/ and /z/ endings in plurals.

● Play the tape and ask the students to listen and repeat the plurals.

Answers
/z/: apples, bananas, eggs, onions, potatoes, tomatoes
/s/: carrots, grapes

2 Aim: to focus on the unstressed *and.*

● Explain to the students that there are many expressions based on two nouns joined by *and.* The *and* in these phrases is pronounced /n/.

● Play the tape and ask the students to repeat the phrases.

LISTENING AND SPEAKING

1 Aim: to prepare for listening.

● Ask the students to read the statements and to say if they are true for their country.

2 Aim: to practise listening for main ideas.

● This activity provides the opportunity for some cross-cultural comparison of eating habits.

● Play the tape and ask the students to tick the true statements and put a cross by the false ones.

	Britain
We eat eggs and bacon for breakfast.	✗
There's always meat and vegetables at the main meal.	✗
We always drink wine at lunch and dinner.	✗
We drink tea during the day.	✓
We often eat potatoes with our main meal.	✓
Many people don't eat meat.	✓

3 Aim: to check comprehension; to practise speaking.

● Play the tape again and ask the students to check their answers in pairs.

● Ask the whole class to check their answers. Ask them if they think eating habits in Britain are similar to those in their countries. If they are different, can they say what the differences are?

3 Look at the words in the vocabulary box. Write *C* for countable or *U* for uncountable.

apple C, ...

4 Write *a* or *some*.

1 book 2 money 3 tea 4 fruit
5 meat 6 pen 7 sandwich
8 chair 9 newspaper 10 coffee

5 Work in pairs and act out the conversation in *Vocabulary and listening*, activity 4.

6 Work in pairs. Make a list of the shopping you need to buy.

Now act out the conversation again with your own shopping list.

SOUNDS

1 Listen and repeat these words. Is the underlined sound /s/ or /z/?

apples bananas carrots
eggs grapes onions potatoes
tomatoes

2 Listen and repeat these phrases. Notice how *and* is pronounced /ən/.

fish and chips
salt and pepper
bread and butter
apples and pears
oranges and lemons
milk and sugar
meat and vegetables

LISTENING AND SPEAKING

1 Read these statements about food and drink. Say if they're true or false for your country.

	Your country	Britain
We eat eggs and bacon for breakfast.		
There's always meat and vegetables at the main meal.		
We always drink wine at lunch and dinner.		
We drink tea during the day.		
We often eat potatoes with our main meal.		
Many people don't eat meat.		

2 Listen to Lisa and find out if the statements in 1 are true or false for Britain. Complete the chart as you listen.

3 Work in pairs and check your answers.

Now listen again and check.

18 I was born in England

Past simple (2): regular verbs; *have*

VOCABULARY AND READING

1 Tick (✓) the verbs you can use to describe events in your life.

live learn work decide play
start receive appear marry

2 You're going to read about the life of the musician and singer Sting. Which of these words do you expect to see?

hit record teenager
rhythm and blues album
song film group award
politician teacher

3 Now read the passage and see if you guessed correctly in 2.

4 Read the passage and match the two parts of the sentences.

1 He was born in Wallsend
2 He learned to play the piano and guitar
3 He started work as a teacher
4 They finished playing together in 1984
5 He received awards for several songs

a but then created *The Police* in 1978.
b when he was a child.
c and worked as a teacher in Newcastle.
d then he appeared in three films.
e then Sting started to sing on his own.

Sting was born in Wallsend, England in 1952, and lived there for over 20 years. As a child he learned to play the piano and guitar. He worked as a teacher in Newcastle but then in 1978 he decided to create the group *The Police* with Stewart Copeland and Andy Summers. They had hits with *Message in a bottle* in 1979 and *Every breath you take* in 1983. *The Police* played together until 1984. Then Sting started to sing on his own and received awards for several songs. He appeared in the films *Brimstone and Treacle* in 1982, *Dune* in 1984 and *Plenty* in 1985. He married Trudy Styler in 1986.

GRAMMAR

Past simple (2): regular verbs

You use the past simple to talk about an action or event in the past which is finished.

*They **finished** playing together in 1984.*

You form the past simple of most regular verbs by adding *-ed* to the infinitive.

learn – learned *work – worked*

You add *-d* to verbs ending in *-e*.

live – lived

The form is the same for all persons.

I *you* *he/she/it* *we* *they*	*lived*	*in London.*

(For information about other regular endings, see Grammar review, page 108.)

Have

The past simple of *have* is *had*. The form is the same for all persons.

*They **had** hits with Message in a bottle.*

18

GENERAL COMMENTS

Sting and Whitney Houston

If the students haven't heard of Sting and Whitney Houston, you will need to explain that they have both been well-known in many countries as singers. Both have been chosen in an attempt to find uncontroversial figures in the world of popular music.

Jigsaw listening

In the *Listening and speaking* section of this lesson, there is an activity which is based on a technique known as jigsaw listening with one tape recorder. The original jigsaw listening involved three or four cassette players and a listening passage divided into three or four parts and re-recorded on to separate cassettes. This version involves just one tape recorder and a task which directs the students to answer different questions as they listen to the same cassette.

VOCABULARY AND READING

1 Aim: to present the words in the vocabulary box.

- Explain that this lesson concerns short biographies of two people. Ask the students to look at the words in the vocabulary box and to use them in sentences to describe important events in their own lives.
- You may like to give the students a few personal examples to help them.
- Ask the students to talk about the life events of a famous person in their country.

2 Aim: to pre-teach some new vocabulary.

- Most of these words occur in the biography of Sting. Check the students understand what they mean. If they remain ignorant of the meaning, they are likely to find the passage quite difficult.
- Find out how many students have heard of Sting. Does anyone know his music? Does anyone like it?

3 Aim: to check the answers to 2.

- Ask the students to read the passage.

Answers
Words in the text: song, film, group, award, teacher

4 Aim: to practise reading for specific information.

- Ask the students to re-read the passage. They can help each other with any difficult vocabulary, but try not to explain too many words yourself.
- Ask the students to match the two parts of the sentences.

Answers
1 c 2 b 3 a 4 e 5 d

GRAMMAR

1 Aim: to focus on the past simple form of regular verbs.

- Ask the students to read the information in the grammar box and then to do the exercises.
- Ask the students to read the passage again and to write down all the regular verbs in the past simple.

Answers
lived, learned, worked, decided, played, started, received, appeared, married

2 Aim: to focus on the past simple form of regular verbs.

- Ask the students to decide what the past simple of these verbs is. Stress that they are regular verbs and that the past simple form is predictable.

Answers
died, stayed, looked, liked, talked, visited, wanted, finished, opened, closed, watched

3 Aim: to practise using the past simple form of regular verbs.

- Do this activity orally with the whole class.

Answers
1 visited 2 watched 3 talked
4 finished 5 died 6 liked

SOUNDS

1 Aim: to focus on the pronunciation of past simple endings.

- Ask the students to listen and notice the pronunciation of the endings.
- Play the tape.

2 Aim: to practise the pronunciation of past simple endings.

- Ask the students to put the words into three columns. You may like to help them by saying the words aloud yourself once or twice, although without pausing.
- Play the tape and ask the students to check their answers. As they listen, they should repeat the words.

Answers
/t/: finished, watched, talked
/d/: learned, closed, died, stayed
/ɪd/: started, visited, wanted

LISTENING AND SPEAKING

1 Aim: to practise listening for specific information.

- Explain to the students that they are all going to listen to the same passage, but will answer different questions.
- Play the tape. Ask the students to work in pairs and follow the Communication activity questions.
- There is no need to check their answers to the questions as they will have a chance to do this in activity 2.

2 Aim: to practise speaking; to check comprehension.

- The students will have listened to the same tape but will have answered different questions. The main point of the activity is in this stage, where they have information separately which will allow them to complete the chart together.

Answers
Whitney Houston
Born: New Jersey, 1963
Started singing: when she was 11, later with blues singers such as Chaka Khan and Lou Rawls.
First hit: *Whitney Houston* in 1985
Appeared in *The Bodyguard*: with Kevin Costner in 1992
Number of copies sold of first two albums: 10 million
I will always love you: 1993

3 Aim: to check comprehension.

- Ask the students to match the two parts of the sentences.

Answers
1 d 2 a 3 b 4 c 5 e

4 Aim: to practise speaking.

- Ask the students to work together and to prepare a short biography of someone famous from their own country.
- You may like to ask your students to do some research for this activity for homework.

1 Read the passage again and find the past simple tense of the verbs in the box in *Vocabulary and reading* activity 1.

live – lived

2 What is the past simple tense of the following regular verbs?

die stay look like talk visit want finish open close watch

3 Complete the sentences with suitable verbs from 2.

1 I _____ the USA in 1996.
2 He _____ the football match on television in the evening.
3 He _____ to her for a long time.
4 The film _____ at half past ten.
5 Shakespeare _____ in 1616.
6 I _____ *Message in a bottle* very much.

SOUNDS

1 Listen to the *-ed* endings of these past simple verbs. Notice how /ɪd/ adds another syllable to the verbs.

/t/	/d/	/ɪd/
worked	lived	decided

2 Put these verbs in the correct columns above.

started learned visited closed died finished wanted stayed watched talked

Now listen and check. As you listen, say the words aloud.

LISTENING AND SPEAKING

1 Work in pairs. You're going to listen to a passage about the singer Whitney Houston.

Student A: Turn to Communication activity 3 on page 99.
Student B: Turn to Communication activity 23 on page 104.

WHITNEY HOUSTON stars as Rachel Marron in Warner Bros.' romantic thriller THE BODYGUARD. Also starring KEVIN COSTNER, THE BODYGUARD is a Tig Production produced by Lawrence Kasdan, Jim Wilson and Kevin Costner, and directed by Mick Jackson. Released in the UK by Warner Bros.

2 Work together and complete the chart with as much detail as possible.

Whitney Houston

Born: ....

Started singing: ....

First hit: ....

Appeared in *The Bodyguard*: ....

Number of copies sold of first two albums: ....

I will always love you: ....

3 Match the two parts of the sentences.

1 She was 11 years old
2 She was 22 years old
3 She appeared in *The Bodyguard*
4 She worked with other singers
5 She received a Grammy Award in 1985

a when she had her first hit album.
b then in the following year she had a hit with a song from the film.
c before she had a hit with her first album.
d when she started singing.
e and in the next year was the first pop singer to sell 10 million copies.

4 Work in pairs. Think of a famous singer or musician in your country. You can use an encyclopaedia. Talk about:

– where he/she was born
– when he/she was born
– where he/she lived as a child
– how long he/she lived there
– when he/she started his/her career
– what he/she did and when

19 What's he like?

Describing people (1)

VOCABULARY

1 Work in pairs. Match the words in the box with the photos.

curly straight elderly young long short tall

2 Put the following adjectives under the following headings: *height, size, hair, age, general impression.*

attractive curly dark elderly fair straight good-looking long medium height young well-built small tall middle-aged old slim pretty short

height:
medium-height . . .

3 You can use the nouns in the box to describe physical features.

beard glasses hair moustache eyes

Look at the photos. Find someone who has:

- short hair
- fair hair
- a beard
- glasses
- a moustache

FUNCTIONS

Describing people (1)

You use the following expressions to describe people.

What's he like?
He's middle-aged and good-looking.
What does he look like?
He's quite tall, with dark hair.
He looks like his mother.
He's got a moustache. She's got glasses.
How tall is he?
He's one metre seventy-five.
How old is he?
He's twenty.

You can also use *What's he/she like?* to ask about someone's character.

What's she like? *She's very nice.*

1 Look at the photos. Listen and decide who the speaker is describing.

2 Choose one of the people in the photos. Write words to describe him or her.

Now work in pairs. Show your partner the list of words. Can you guess who your partner is describing?

3 Work in pairs. Describe the following people to each other.

- one of your parents or grandparents
- a brother or sister
- your best friend
- one of your parents' friends
- someone in the class

WRITING

1 Read the letter. Does Nick know Pat?

33, Highclere Road
Basingstoke
England
20 May

Dear Pat,

Thanks for your letter. It's very kind of you to let me stay with you and to meet me at the airport.

I arrive at 22.30 on Thursday 22 June. I'm twenty and I've got dark curly hair. I'm quite short, about one metre seventy and I'm quite well-built. If you have time, send me a description of yourself so I can recognise you.

See you at the airport.

Best wishes,

Nick

19

GENERAL COMMENTS

Cultural sensitivities

There are some cultures which are reluctant to talk about the physical appearance of people. Even if your students do not belong to such a culture, the subject may need to be treated sensitively. For example, words which may be perceived as unkind such as *fat* are not presented in this lesson, although you may think they are suitable to include.

VOCABULARY

1 Aim: to present the words in the vocabulary box.

- Ask the students to match the words with the photos.
- Check everyone knows what the words mean.

Answers
photo 1: curly, elderly
photo 2: short
photo 3: straight, long, young,
photo 4: tall

2 Aim: to present the words in the vocabulary box.

- The words in this box describe more subjective impressions of appearance and are therefore more difficult to illustrate. You may need to think of well-known people to illustrate the meaning of some of the words.
- Ask the students to put the adjectives under the various headings.

Answers
height: medium height, short, small, tall
size: slim, well-built
hair: curly, dark, fair, long, short, straight
age: elderly, middle-aged, old, young
general appearance: attractive, good-looking, pretty

3 Aim: to present the words in the vocabulary box.

- Ask the students to use the words in the box to describe people in the photos or in the class.

Answers
photo 1, 2 and 4: short hair
photo 2: beard, moustache
photo 3: fair hair, glasses

FUNCTIONS

1 Aim: to present the language of describing people.

- Ask the students to read the information in the functions box and then to do the exercises.
- Explain to the students that they are going to hear someone describing the people in the photos. Ask them to listen and decide who is being described.
- Play the tape.

Answers
Conversation 1: 1
Conversation 2: 4
Conversation 3: 2
Conversation 4: 3

2 Aim: to practise describing people.

- Ask the students to work on their own and to write a brief description of someone in the photos.
- When they are ready, ask the students to exchange their descriptions. The other student should guess who his/her partner has described.

3 Aim: to practise describing people.

- Ask the students to work in pairs and to describe someone in their family.
- Ask two or three students to give their descriptions to the whole class.

WRITING

1 Aim: to practise describing people; to prepare for writing; to practise reading for main ideas.

- Ask the students to read the letter and to decide if Nick knows Pat.

Answer
No, he doesn't

2 Aim: to practise describing people.

- Explain that Nick is one of the people in the photos. Can anyone see who it is?

Answer
2

3 Aim: to practise describing people; to practise writing.

- Ask the students to write a letter describing themselves to someone they don't know.
- Make sure they use the letter in activity 1 as a model. Check that the layout is appropriate. You may like to discuss the content of each paragraph and why the writer has started a new paragraph on each occasion.

LISTENING AND SPEAKING

1 Aim: to practise speaking; to prepare for listening.

- This activity focuses on stereotypical physical appearances of people from different countries.
- Ask the students if there is any physical feature which is typical of people from their countries, such as hair colour or eye colour.
- Ask the students to read the statements and decide if they are true or false for people from their countries.
- Discuss the answers with the whole class.

2 Aim: to practise listening.

- Play the tape and ask the students to say if the statements are true or false for Kevin.

Answers
1 False 2 False 3 True 4 False

3 Aim: to practise speaking; to practise listening for specific information.

- Ask the students to remember what Kevin said in greater detail.
- Encourage the students to try to reconstitute as much of what Kevin said as possible.
- Play the tape for students to check their answers.

Answers
1 b 2 c 3 a 4 c

2 Decide which of the people in the photos is Nick.

3 Write a letter to Nick describing yourself. Describe:

– your age – your hair – any physical features
– your height – your size

Make sure you put your address and the date on the letter.

LISTENING AND SPEAKING

1 Look at these statements and decide if they are mostly true or false for your country.

1 People are tall when they're over one metre sixty centimetres.
2 People are old when they're over sixty.
3 People are middle-aged when they're over forty.
4 People are usually quite well-built.

2 Listen to Kevin, who is Irish, talking about the statements. Are they true or false for his country?

3 Work in pairs. What did Kevin say about the statements?

1 People are tall when they're over:
a one metre sixty centimetres
b one metre eighty centimetres c two metres
2 People are old when they're:
a fifty b sixty c seventy
3 People are middle-aged when they are:
a forty b fifty c sixty
4 People are usually:
a well-built b slim c well-built or slim

Can you remember any other details?

Now listen again and check.

20 A grand tour

Past simple (3): irregular verbs; *yes/no* questions and short answers

SOUNDS

1 Here are the names of some cities in Europe. Match them with the English name of the country they're in.

Budapest Zurich London Paris Venice Vienna

Austria England France Hungary Italy Switzerland

2 [cassette] Listen and check your answers to 1. As you listen, say the words aloud.

3 What's the English name for your country or city? Is it the same as in your language?

VOCABULARY AND LISTENING

1 Match the verbs in the box with the phrases below.

buy do find fly go have listen lose make read stay take visit watch write

to the Opera	some souvenirs
a museum	a meal
some shopping	a cheap hotel
your wallet	home
to a concert	friends
the newspaper	some postcards
with friends	a tram
the football	

2 Work in pairs and look at the map of Europe. Find two or three places you want to visit. Can you think of things you want to do there?

3 [cassette] Listen to Mary and Bill from the USA who went on a tour around Europe. Follow their route on the map.

4 Before Mary and Bill started their tour, they made a list of things they wanted to do. Tick (✓) the things they did.

buy some souvenirs	☐	have a steam bath	☐
visit the museums and galleries	☐	write some postcards	☐
fly home	☐	go to the Opera	☐
do some shopping	☐	take a tram	☐
find a cheap hotel	☐	lose a wallet	☐
stay with friends	☐	have a meal in a smart restaurant	☐
meet some New Yorkers	☐	read the English newspapers	☐
make friends with local people	☐	relax in the parks	☐
listen to a concert	☐	see the Queen	☐

[cassette] Listen again and check.

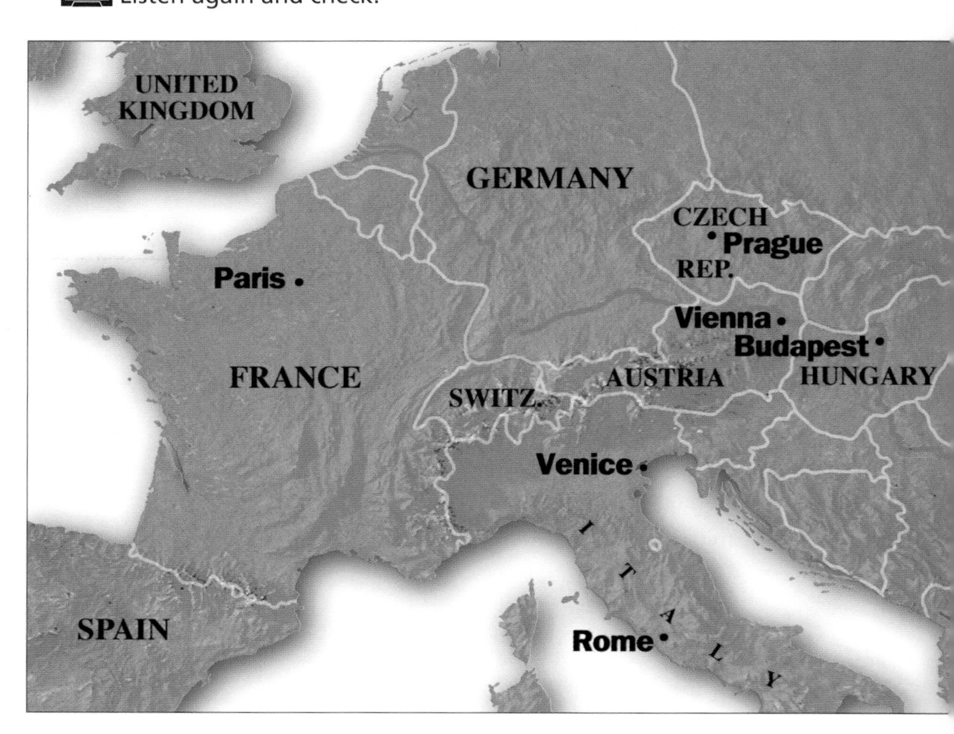

20

GENERAL COMMENTS

Towns and countries

It is not just the names of countries in English which may be different from your students' own language. The names of towns may be different as well. This lesson begins by focusing on these differences.

Irregular verbs

Explain to the students that the past simple form of irregular verbs has to be learnt as a vocabulary item in its own right. Even though some irregular verbs follow certain patterns, there are no generative rules about these verbs – otherwise they wouldn't be irregular!

Making your own recordings

If you're interested in collecting your own recorded material, especially on the topic of travel, this lesson presents a model for you to follow. Ask a colleague or a native-English speaker to talk into a microphone about a journey they have made around a continent, a country or a region, and process the recording in the same way as the sequence of activities shown in *Vocabulary and listening*.

SOUNDS

1 Aim: to present the English pronunciation of some European towns and countries.

- Ask the students to match the towns and the countries.

Answers
Budapest - Hungary, Zurich - Switzerland, London - England, Paris - France, Venice - Italy, Vienna - Austria

2 Aim: to check the pronunciation of towns and countries.

- Play the tape and ask the students to check their answers to 1. Ask them to say the words aloud.

3 Aim: to check the pronunciation of towns and countries.

- Check everyone knows what the English name is for their own town and country.

- You may like to do a similar activity to 1 with the English form of towns and countries in your region of the world.

VOCABULARY AND LISTENING

1 Aim: to present the words in the vocabulary box.

- Once again, this is an activity which focuses on words which go together. Ask the students to match the verbs and the noun phrases.

Possible answers
buy some souvenirs, do some shopping, find a cheap hotel, fly home, go to the Opera, have a meal, listen to a concert, lose your wallet, make friends, read the newspaper, stay with friends, take a tram, visit a museum, watch the football, write some postcards

- There are other possible answers, but these are the ones you'll hear in the listening passage.

2 Aim: to prepare for listening.

- Ask the students to look at the map of Europe and to find places they'd like to visit. Ask them to work in pairs and then ask them to tell the rest of the class where they'd like to go and why.

3 Aim: to listen for main ideas.

- Play the tape and ask the students to follow Mary and Bill's route around Europe.

Answer
Paris, Venice, Vienna, Budapest, Paris, then back to New York.

4 Aim: to check comprehension; to prepare to practise the target structures.

- Ask the students to tick the things Mary and Bill did on their tour.

- Play the tape. Students may like to check their answers in pairs.

Answers

buy some souvenirs	✓
visit the museums and galleries	✓
fly home	✓
do some shopping	✓
find a cheap hotel	✓
stay with friends	
meet some New Yorkers	✓
make friends with local people	
listen to a concert	
have a steam bath	✓
write some postcards	
go to the Opera	✓
take a tram	
lose a wallet	✓
have a meal in a smart restaurant	
read the English newspapers	
relax in the parks	
see the Queen	
watch the football	

GRAMMAR

1 Aim: to present the past simple form of irregular verbs.

- Ask the students to read the information in the grammar box and then to do the exercises.
- You may like to do this activity orally with the whole class.

Answers
become - became, buy - bought, come - came, find - found, fly - flew, have - had, hear - heard, get - got, give - gave, go - went, leave - left, lose - lost, make - made, read - read /red/, say - said, sell - sold, send - sent, spend - spent, wake - woke, win - won, write - wrote

2 Aim: to focus on the auxiliary in past simple questions.

- Ask the students to complete the sentences with the correct auxiliary.

Answers
1 Did 2 Were 3 Did 4 Was 5 Did 6 Did

3 Aim: to focus on the past simple form of irregular verbs.

- Ask the students to do this activity in pairs and in writing.

Answers
They went to Paris, and did some shopping.
They went to Venice, and met some New Yorkers.
They went to Venice and did a lot of sightseeing.
They went to Vienna and found a cheap hotel.
They went to Vienna and went to the Opera.
They went to Budapest and Mary bought some souvenirs, and Bill had a steam bath.
They went to Budapest and Bill lost his wallet.
They flew back to Paris and then flew to New York.

4 Aim: to focus on short answers.

- Ask the students to do this orally.

Answers
1 Yes, he did. 2 No, they didn't. 3 No, they didn't.
4 No, they didn't. 5 Yes, they did. 6 Yes, he did.
7 Yes, they did. 8 No, they didn't.

SPEAKING

1 Aim: to practise speaking; to practise using the target structures.

- Ask the students to go round talking to other students and asking about their holidays.
- Ask the students to write down the names of the people they find in the *Find someone who* ... activity.
- Ask several students to report back to the class on who they found had done the things mentioned.

2 Aim: to practise speaking; to practise using the target structures.

- Ask the students to work in pairs and to talk about their holidays.
- You may like to ask the students to write a short paragraph about their holidays, using the prompts in this activity.

GRAMMAR

Past simple (3): irregular verbs

Many verbs have an irregular past simple form.

buy – bought *write – wrote*
go – went

For a list of irregular verbs, see page 114.

Yes/no questions and short answers

You form *yes/no* questions with *did*.

Did they stay with friends in Paris?
Yes, they did. No, they didn't.

1 Match the infinitives with their irregular past tenses.

become buy come find fly have hear get give go leave lose make read say sell send spend wake win write

had came left went became wrote got gave won lost found bought sold read spent woke said flew made sent heard

2 Complete the sentences with *did, was* or *were*.

1 _____ they go to the Eiffel Tower?
2 _____ they born in the USA?
3 _____ they have a good time?
4 _____ it sunny in France?
5 _____ they do some shopping in Paris?
6 _____ they make friends with anyone?

3 Work in pairs. Look at your answers to *Vocabulary and listening* activities 3 and 4 and say where Mary and Bill went and what they did.

They went to Paris and did some shopping.

4 Give short answers to the questions.

1 Did Bill have a steam bath in Budapest?
2 Did they see the Queen in London?
3 Did they watch the football in Paris?
4 Did they take a tram in Vienna?
5 Did they find a cheap hotel in Vienna?
6 Did Bill lose his wallet in Budapest?
7 Did they make friends in Venice?
8 Did they write some postcards in Budapest?

1 Yes, he did.

SPEAKING

1 Ask other students about their last holiday. Find someone who:

took a plane
stayed with friends
did some sightseeing
lost their passport
stayed in a hotel
went to Britain
had a meal in a restaurant
stayed in a tent
made some new friends

Did you take a plane?
Yes, I did.
Did you stay with friends?
No, I didn't.

2 Work in pairs. Tell each other about a holiday you had. Talk about:

– where you went
– what you did
– where you stayed
– how long you stayed there
– what the weather was like
– when you went home

Progress check 16–20

VOCABULARY

1 When you write down new words, write down other words which you associate with them. They don't have to be the same part of speech.

musician – *song, guitar, rhythm and blues*
fly – *plane, airport*

Match the words in box A with the associated words in box B. There may be more than one possibility.

A | hit record bread polite marriage attractive restaurant friendly

B | album well-behaved divorced cheerful butter good-looking meal

hit record – album

2 You can use word charts to write down your new words.

TRANSPORT			
Road	**Rail**	**Air**	**Sea**
bicycle, car, bus garage, bus stop	train, underground station	plane airport	ship, boat ferry, port

Write word charts for one of the following. Use a dictionary, if necessary.

Home – (think of rooms and furniture)
Time – (think of days and months)
Places – (think of adjectives to describe a town, facilities etc)

3 It's also useful to write down the other parts of speech of a word.

Noun	**Verb**
writer	write
success	succeed

Use your dictionary to find:
– nouns formed from the following verbs

decide play marry visit start

– nouns formed from the following adjectives

happy lazy attractive young

GRAMMAR

1 Complete the sentences with *some* or *any.*

1 I'd like _____ apples, please.
2 Do we need to get _____ rice?
3 They want _____ tea.
4 I haven't got _____ money on me.
5 Let's buy _____ apples.
6 There isn't _____ toast.

2 Write *C* for countable or *U* for uncountable.

water carrot juice egg onion lamb apple milk oil burger cheese pasta potato

water U, ...

3 Write the past simple form of the following verbs.

become work decide buy appear visit have go die start lose land paint read invent send discover

4 Match the questions 1 – 4 with the answers a – d.

1 What's she like?
2 What does she look like?
3 How tall is she?
4 How old is she?

a About one metre seventy.
b She's tall with long, dark hair.
c About twenty.
d She's very pretty.

Progress check 16 - 20

GENERAL COMMENTS

You can work through this Progress check in the order shown, or concentrate on areas which have caused difficulty in Lessons 16 - 20. You can also let the students choose the activities they would like or feel the need to do.

VOCABULARY

1 Aim: to help the students organise their new vocabulary and related words.

● Remind the students that their vocabulary notebooks should not be just lists of words, but a kind of network in which the same word appears several times and is linked to words, categories or topics.

Answers
hit record - album, bread - butter, polite - well-behaved, marriage - divorced, attractive - good-looking, restaurant - meal, friendly - cheerful

2 Aim: to help the students organise their vocabulary with word charts.

● Ask the students to make their own word charts with *home, time* and *places.*

● Do make sure you look at their word charts and make suggestions, but it is advisable not to give them marks. This type of activity is designed to increase the learner's independence.

3 Aim: to help the students organise their vocabulary by writing down other parts of speech.

● It may be necessary to point out that a word in its different part of speech may be used in very different contexts.

Answers
Nouns formed from the verbs:
decision, play, player, marriage, visitor, starter
Nouns formed from the adjectives:
happiness, laziness, attraction, youth

GRAMMAR

1 Aim: to revise *some* and *any*.

Answers
1 some 2 any 3 some 4 any 5 some 6 any

2 Aim: to revise countable and uncountable nouns.

Answers
Countable: carrot, egg, onion, apple, burger, potato
Uncountable: water, juice, lamb, milk, oil, cheese, pasta

3 Aim: to revise the past simple of regular and irregular verbs.

Answers
became, worked, decided, bought, appeared, visited, had, went, died, started, lost, landed, painted, read, invented, sent, discovered

4 Aim: to revise the language for describing people.

Answers
1 d or b 2 b or d 3 a 4 c

5 Aim: to revise the language for describing people.

- Ask the students to write a description of either one or both of the people in the drawings. Suggest that they answer the questions in activity 4.

6 Aim: to revise the use of the auxiliaries in the past simple.

Answers
1 Was 2 Was 3 Did 4 Were 5 Did 6 Did

7 Aim: to revise short answers in the past simple.

Answers
1 No, he wasn't.
2 Yes, there was.
3 Yes, I did.
4 No, there weren't.
5 No, they didn't.
6 Yes, she did.

SOUNDS

1 Aim: to focus on /ɔ/, /ɒ/ and /ɜː/

- It may come as a surprise to the students that the spelling of English words is only an approximate guide to their pronunciation.
- Ask the students to group the words according to the underlined sound.

Answers
There are three groups:
/ɔ/: award, bought, divorced, morning
/ɒ/: got, lost, was
/ɜː/: learned, return, served, third, word

- Play the tape and ask the students to say the words aloud.

2 Aim: to focus on silent letters.

- Ask the students to say the words in this activity and to decide which letters are not pronounced.

Answers
ca~~r~~d, dau~~gh~~ter, cu~~p~~board, bou~~gh~~t, i~~s~~land, lam~~b~~

- Play the tape and ask the students to say the words aloud.

3 Aim: to focus on stressed syllables in words.

- Ask the students to say the words aloud and match the words with the stress patterns.

Answers
■▪ teacher, evening
▪■ award, appear, receive
▪■▪ unhappy, eleventh, tomato, banana

- Play the tape and ask the students to check their answers and say the words aloud.

4 Aim: to focus on the tone of voice.

- Play the tape and ask the students to listen to the sentences in *Grammar* activity 6. The students should decide if the speaker sounds bored or interested.

Answers
1 bored 2 interested 3 interested
4 bored 5 bored 6 interested

- Ask several students to say the sentences aloud to the whole class. Make sure they sound lively and interested.

SPEAKING

1 Aim: to practise speaking; to revise the present and past simple tenses.

- You may need to model this game yourself. If so, put the students into groups of three or four. Think of someone famous, alive or dead. Encourage the students to ask you *yes/no* questions, and make sure you give short answers. Count the number of questions each group asks before they guess the answer. Explain that the group with the lowest number of questions wins.
- Now ask the students to play this game themselves with one student from each group thinking of a famous person and the other students asking questions. Ask them to keep score of how many questions they ask.

5 Write a description of the people in the picture.

6 Complete the questions with *did, was* or *were*.

1 _____ Jonas born in Vienna?
2 _____ there anything to eat at the party?
3 _____ you see the football match last night?
4 _____ there lots of people on the bus?
5 _____ they arrive on time?
6 _____ she get home safely?

7 Here are the answers to the questions in 6. Write short answers.

1 No 2 Yes 3 Yes 4 No 5 No 6 Yes

1 No, he wasn't.

SOUNDS

1 Group the words which have the same underlined sound.

aw<u>ar</u>d b<u>ough</u>t div<u>or</u>ced g<u>o</u>t l<u>ear</u>ned l<u>o</u>st m<u>or</u>ning ret<u>ur</u>n s<u>er</u>ved th<u>ir</u>d w<u>a</u>s w<u>or</u>d

How many groups are there?

Now listen and check. As you listen, say the words aloud.

2 Many words have letters which you don't pronounce.

ei~~gh~~t wo~~r~~d lis~~t~~en

Cross out the silent letters in these words.

card daughter cupboard bought island lamb

Now listen and check. As you listen, say the words aloud.

3 Match the words with the stress patterns.

□ ▫ ▫ □ ▫ □ ▫

award appear teacher receive unhappy eleventh evening tomato banana

Listen and check. As you listen, say the words aloud.

4 Listen to the questions in *Grammar* activity 6. Put a tick (✓) if you think the speaker sounds interested.

Now say the questions aloud. Try to sound interested.

SPEAKING

Work in groups of three or four. Play *Twenty Questions*.

Student A: Think of a person. Answer Student, B, C and D's questions with *Yes* and *No* until they find out who the person is.

Student B, C and D: Student A is thinking of a person. In turn, ask questions to find out who the person is. You can only ask twenty questions in all.

1 Is it a man? Yes.
2 Is he alive? No.
3 Was he a singer? Yes.

Change round when you're ready.

21 *Mystery*

Past simple (4): negatives; *wh-* questions

Agatha Christie was the most successful writer of detective stories of all time. People all over the world read her stories of Hercule Poirot and Miss Marple. But when she died in 1976 there was a final mystery: why did she disappear for eleven days in December 1926?

Agatha Christie was born in September 1890. She lived with her family in Devon, England. In 1914 she married Colonel Archibald Christie. She wrote her first detective story in 1920 and soon she was very successful.

But Agatha Christie didn't have a happy marriage. On a cold night in December 1926 she left home in her car. The following morning, the police found the empty car but there was no sign of Agatha Christie. Two days later, they told the newspapers that they didn't know where she was. Everyone thought she was dead.

But 250 miles away in Yorkshire, a waiter in a hotel saw a guest who looked like Agatha Christie, and he told the police. Eleven days after her disappearance, her husband found her again in the hotel dining room.

The couple were soon divorced. She married Sir Max Mallowan, an archaeologist, in 1930 and she continued to write her mysteries. But she didn't explain what happened in 1926. Did she want to kill herself? Did she want to show her husband that she didn't love him? Did she hope to sell more books?

Over the years, Agatha Christie wrote more than 80 mysteries and sold over 300 million books. But she didn't tell anyone why she disappeared in December 1926.

Why did Agatha Christie disappear?

PENGUIN BOOKS
MURDER ON THE ORIENT EXPRESS
AGATHA CHRISTIE
MYSTERY AND CRIME
1/6

21

GENERAL COMMENTS

Agatha Christie

You may like to mention that Agatha Christie created the detectives Hercule Poirot and Miss Marple, and some of her most famous books have been turned into films, such as *Murder on the Orient Express* and *Death on the Nile.*

VOCABULARY AND READING

1 Aim: to prepare for reading.

- Ask the students if they have heard of Agatha Christie. Ask them who she was, what nationality she was, what she did, when she lived, who were her famous characters.

2 Aim: to prepare for reading; to pre-teach difficult words.

- Tell the students that the article they are going to read is, in fact, not about one of Agatha Christie's characters, but about the writer herself.

- Ask the students to work in pairs. They should read the words and check they understand them. Ask them to prepare a brief oral description of what happened to Agatha Christie.

- Ask several pairs to tell the whole class what they think happened to Agatha Christie. Of course it won't matter if they do not make an accurate description, but the process of predicting the contents of the article will be excellent preparation for reading.

3 Aim: to practise reading for main ideas.

- Ask the students to read the article and to choose the correct answer to the question in the title.

> **Answer**
> 2

- Try not to explain too many words for the moment, as this will detract from the aim of the activity, which is to form a general impression of the whole passage.

4 Aim: to practise reading for specific information.

- Ask the students to look at the questions and to try and answer them from what they remember of the first reading.
- Ask the students to check their answers by reading the article again carefully.

> **Answers**
> 1 False 2 False 3 False 4 True
> 5 False 6 False 7 True 8 False

5 Aim: to practise speaking.

- Ask the students to work in pairs and to ask and answer the questions in the Communication activities.

GRAMMAR

1 Aim: to practise forming the past simple negative.

- Ask the students to read the information in the grammar box and then to do the exercises.
- Ask the students to do this exercise in writing.

> **Answers**
> 1 She didn't marry when she was fifteen. She married when she was twenty-four.
> 2 The police didn't find her in the car. The car was empty.
> 3 The police didn't say she was dead. Everyone thought she was dead.
> 5 Her husband didn't find her at home. He found her in a hotel 250 miles away.
> 6 She didn't marry a crime writer. She married an archaeologist.
> 8 She didn't die at the age of seventy-six. She died at the age of eighty-six.

2 Aim: to practise writing questions.

- You may like to do this activity orally with the whole class. Make sure the auxiliary *did* goes in the right place.
- Make it clear that there may be more than one possible question.

> **Possible questions**
> 1 What did Agatha Christie write?
> 2 When was she born?
> 3 Where did she live?
> 4 Who did she marry?
> 5 When did she disappear?
> 6 Where did her husband find her?
> 7 When did he find her?
> 8 What did Sir Max Mallowan do?

3 Aim: to practise writing questions and negatives.

- Ask the students to work in pairs and to continue to practise writing statements and negatives.

SOUNDS

1 Aim: to focus on the unstressed pronunciation of auxiliary verbs.

- Play the tape and ask the students to listen to how you don't usually stress the auxiliary verbs in questions.
- Ask the students to say the sentences aloud.

WRITING AND SPEAKING

1 Aim: to practise writing; to practise using the target structures.

- Ask students to write notes about their lives in preparation for their autobiographies.
- Ask the students to write simple sentences based on their notes.

2 Aim: to practise writing; to practise using the target structures.

- Ask the students to exchange their autobiographies and to write extra questions about their partner's autobiographies.

3 Aim: to practise writing; to practise using the target structures.

- Ask the students to re-write their autobiographies with their answers to the questions their partners asked.
- You may like to ask the students to do this last stage for homework.

VOCABULARY AND READING

1 Work in pairs. You're going to read an article about Agatha Christie. Do you know who she was? Talk about anything you know about her. Think about what you'd like to find out about her.

2 Work in pairs. Look at the title of the article and the words in the box. What do you think happened?

successful writer detective story mystery marriage unhappy left home disappear waiter hotel guest husband find divorced kill tell

3 Read the article and choose the correct answer to the question in the title.

1 Because she was unhappy.
2 We don't know.
3 Because she wanted to kill herself.

Did you guess correctly in 2?

4 Put a tick (✓) by the statements which are true.

1 Agatha Christie married when she was fifteen.
2 The police found her in the car.
3 The police said that she was dead.
4 A hotel waiter recognised her.
5 Her husband found her at home.
6 She married a crime writer.
7 She wrote over eighty detective stories.
8 She died at the age of seventy-six.

5 Work in pairs.

Student A: Turn to Communication activity 8 on page 100.

Student B: Turn to Communication activity 15 on page 102.

GRAMMAR

> Past simple (4): negatives
>
> **You form the negative for all verbs except *be* with *didn't* + infinitive without *to*.**
>
> *She* ***didn't have*** *a happy marriage.* *She* ***didn't explain*** *her disappearance.*
>
> *Wh-* questions
>
> **You form *Wh-* questions with a *Wh-* question word *(who, what, when, why)* + *did* + infinitive without *to*.**
>
> ***What did*** *Agatha Christie* ***do?*** ***Why did*** *she* ***disappear?***
>
> *Where* ***was*** *she born?*

1 Correct the false statements in *Vocabulary and reading* activity 4.

She didn't marry when she was fifteen. She married when she was twenty-four.

2 Here are some answers about Agatha Christie. Write the questions.

1 She wrote detective stories.
2 In September 1890.
3 In Devon.
4 Colonel Archibald Christie.
5 In December 1926.
6 In a hotel in Yorkshire.
7 Eleven days after her disappearance.
8 He was an archaeologist.

3 Turn to Communication activity 6 on page 100.

SOUNDS

Notice how you don't usually stress auxiliary verbs.

Listen to the pronunciation of the auxiliary verbs in the questions you wrote in *Grammar* activity 2.

Now say the sentences aloud.

WRITING AND SPEAKING

1 Write a short autobiography. Say:

– where you were born
– where you lived
– when you started school
– if there were any special events in your life

I was born in 1982. I lived with my family in Malaga. Then we moved to Seville.

2 Work in pairs and exchange your autobiographies. Write extra questions about your partner's autobiography.

How long did you live in Malaga? When did you move to Seville?

3 Read your partner's questions. Rewrite your autobiography and include the answers.

I was born in 1982 in Malaga, and I lived there with my family for ten years. Then we moved to Seville in ...

22 Dates

Past simple (5); expressions of time

VOCABULARY AND SOUNDS

1 Match the words in the box with the numbers below.

eighth eleventh fifth first fourth ninth second seventh sixth tenth third twelfth

1st 2nd 3rd 4th 5th 6th 7th 8th 9th 10th 11th 12th

1st – first

2 Write the words for the following numbers.

13th 17th 20th 21st 22nd 23rd 27th 30th 31st

13th – thirteenth

3 Notice how:

– you write

1st March 13th April

– you say

the first of March the thirteenth of April

Now listen and repeat these dates.

1st March 13th April 23rd September 4th January 30th July 2nd May 20th June 10th October

22

GENERAL COMMENTS

Dates

There are a few extra details about dates which you may want to tell your students.
When you write a date you usually put a comma before the year only when the date is within a sentence.
I was born on 20 June, 1981.
There is some discussion about what we will call the years in the 21st century. At the moment we usually say *in the year 2000*, and rarely *in 2000*. We also talk about *two thousand and ten, two thousand and thirty-three*, and so on, which does not follow the pattern of the years up to the turn of century (*nineteen ten, eighteen thirty-three*). It's difficult to predict whether the exceptional usage will establish itself, or whether the conventional usage will reassert itself.

VOCABULARY AND SOUNDS

1 Aim: to present ordinal numbers.

- Write the ordinal numbers on the board and say them aloud.
- Ask the students to say the words aloud, in chorus and then individually.
- Ask the students to match the words in the box with the numbers below.
- You may like to ask students to do this activity with a partner.

Answers
2nd - second
3rd - third
4th - fourth
5th - fifth
6th - sixth
7th - seventh
8th - eighth
9th - ninth
10th - tenth
11th - eleventh
12th - twelfth

2 Aim: to practise ordinal numbers.

- Ask the students to write the words for the numbers shown.
- You may want to write some extra numbers on the board.

Answers
17th - seventeenth
20th - twentieth
21st - twenty-first
22nd - twenty-second
23rd - twenty-third
27th - twenty-seventh
30th - thirtieth
31st - thirty-first

3 Aim: to present saying and writing dates.

- Ask the students to read the information about the dates.
- Play the tape and ask the students to repeat the dates they hear.
- Continue this activity by writing a few extra dates on the board.

4 Aim: to present the months of the year; to practise saying dates.

- Ask the students to say the dates of the occasions mentioned. You may be able to add to this list.

LISTENING AND SPEAKING

1 Aim: to pre-teach some difficult vocabulary; to prepare for listening.

- Your students may not know some of these words, so it would be useful to pre-teach them before you play the tape.
- Ask the students to match the words with the special days. There may be more than one possible answer.

Answers
an important birthday: present, letter, party, card, forget
passing an exam: driving licence, certificate
Independence Day: party
a wedding day: present, church, reception, certificate, card, ring
an anniversary: present, letter, party, card, forget

2 Aim: to practise listening for main ideas.

- Ask the students to listen and decide which special day the speaker is talking about.
- Try not to explain too much vocabulary if you want to give the students some effective practice in listening to difficult English.
- Play the tape.

Answers
Speaker 1: a wedding day
Speaker 2: passing an exam
Speaker 3: an anniversary

3 Aim: to provide an opportunity for a second listening; to present some expressions of time.

- Ask the students to work in pairs and to try and remember as much as possible about the three passages.
- Play the tape for students to check.

Answers
Speaker 1: five months ago, in August, yesterday evening
Speaker 2: last Thursday, at the end of the year
Speaker 3: in 1987, on the eleventh of December

GRAMMAR

1 Aim: to practise using the expressions of time.

- Ask the students to read the information in the grammar box and then to do the exercises. You may need to explain or translate some of them.
- Ask the students to complete the sentences.
- Ask several students to read out their completed sentences.

2 Aim: to practise using expressions of time.

- Ask the students to work in pairs and to ask and answer the questions.
- Ask several students to tell the whole class what their answers to the questions were.

SPEAKING

1 Aim: to practise speaking; to practise using the target structures.

- Ask the students to work in pairs and write their answers to the questions on a piece of paper.
- When everyone is ready, collect the papers and give them out again to be marked. Make sure the students don't mark their own answer papers.

Answers

1 1989
2 1917
3 1969
4 1945
5 1492
6 from 214BC
7 1939
8 1963
9 1914

2 Aim: to practise speaking and writing; to practise using the target structures.

- Ask the students to transform the statements in their Communication activities into questions.

Answers
Student A:
1 When was the Kobe earthquake?
2 When did Martin Luther King die?
3 Who invented the telephone?
4 How old was Michelangelo when he died?
5 Where was Joseph Stalin born?

Student B:
1 When was the Mexico City earthquake?
2 Who was the first man in space?
3 When did the first American walk on the moon?
4 How many wives did King Henry VIII of England and Wales have?
5 Where was Sigmund Freud born?

- Ask the students to ask and answer the questions in groups and to continue the scoring.

4 Work in pairs. What dates are the following?

New Year's Day your birthday
the national day or an important day in your country yesterday today

January February March April May June July August September October November December

New Year's Day is the first of January.

LISTENING AND SPEAKING

1 Match these words with the special days below.

present church letter party reception
driving licence certificate card forget ring

☐ an important birthday
☐ passing an exam
☐ Independence Day
☐ a wedding day
☐ an anniversary

2 You're going to hear three people talking about one of the special days. Listen and put the number of the speaker by the special day he/she is describing in 1.

3 Work in pairs and check your answers to 2. Which speaker uses the following expressions?

last Thursday yesterday evening five months ago
in 1987 in August from nine to five
on the eleventh of December at the end of the year

Listen again and check.

GRAMMAR

Past simple (5): expressions of time

You can use these expresssions of past time to say when something happened.

last *night/Thursday/August/month/year*
I saw him ***last year.***

in *August/1987*
We got married ***in 1987.***

ago *days/weeks/months ago*
We went on holiday ***three weeks ago.***

yesterday *morning/afternoon/evening*
I met her ***yesterday morning.***

at the end of *the day/month/year*
My birthday is ***at the end of the month.***

on *Monday/the eleventh of December*
My birthday is ***on the eleventh of December.***

1 Complete the sentences with an expression of time in the grammar box.

1 I went to the dentist ______.
2 I did my homework ______.
3 I started learning English ______.
4 I bought someone a present ______.
5 I met my best friend ______.
6 I wrote a postcard ______.

2 Work in pairs. Ask and answer the questions.

1 When did you last buy a new coat?
2 When did you get home yesterday?
3 When did you last go to the cinema?
4 When did you get up this morning?
5 When did you last have a birthday?
6 When did you start this lesson?

1 A year ago.

SPEAKING

1 Work in pairs and answer the questions in the quiz. You score 2 points if you get the right answer, and 1 point if it's very close. Your teacher will decide.

When did ...

1 ... the Berlin Wall come down?
2 ... the Russian Revolution start?
3 ... astronauts first land on the moon?
4 ... the first atom bomb explode?
5 ... Columbus discover America?
6 ... the Chinese build the Great Wall?
7 ... the Second World War start?
8 ... John Kennedy die?
9 ... the First World War start?
10 ... you start learning English?

2 Work in two pairs and continue the quiz in 1.

Pair A: Turn to Communication activity 5 on page 100.
Pair B: Turn to Communication activity 22 on page 104.

What's she wearing?

Describing people (2); present continuous or present simple

VOCABULARY AND LISTENING

1 Look at the people in the photos and say what they're wearing. Use the words in the box to help you.

trousers jeans skirt socks shorts dress tie shirt T-shirt jacket sweater shoes trainers

Turn to Communication activity 13 on page 102 to find out what the other words are.

2 Look at the words in the box again. Find things:

– you wear on your feet
– that men usually wear
– that women usually wear
– you wear when it's hot
– you wear when it's cold

3 Which words can you use to describe the clothes:

– you're wearing at the moment?
– you usually wear?

comfortable fashionable casual smart warm

4 Look at the photos again and use these words to describe the people.

sit down stand smile laugh wear

5 Listen to Jan and find out who the people in the photos are.

6 Work in pairs and check your answers. Complete the chart with as much detail as possible.

Listen again and check.

	Harriet	John	Edward	Louise
What's he/she wearing?				
What's he/she doing?				

23

GENERAL COMMENTS

Present simple and present continuous

These two tenses have been presented in isolation until now, and many students whose mother tongue does not make a distinction between them, may now be confusing them on a regular basis. If your students need more practise on this structure use the extra material in the Practice Book and the Resource Pack.

American and British English

If you have time, this may be a suitable lesson to explore some of the differences between British and American English. There are many differences in the vocabulary of clothes. For example:

British English	American English
(under)pants	shorts
trousers	pants
vest	undershirt
waistcoat	vest
dressing gown	bathrobe
lounge suit	business suit

VOCABULARY AND LISTENING

1 Aim: to present the words in the vocabulary box.

- Ask the students to show they understand the meaning of the words by pointing to items of clothing.
- If there is any confusion, ask the students to turn to the relevant Communication activity for an illustrated explanation of the meaning of all the words.

2 Aim: to check comprehension of the words in the box.

- This categorisation activity is designed to consolidate the learning process. Ask the students to do this activity on their own.
- Ask two or three students to read out their answers to the whole class.

Answers
things you wear on your feet: socks, shoes, trainers
things that men usually wear: trousers, jeans, socks, shorts, tie, shirt, T-shirt, jacket, sweater, shoes, trainers
things that women usually wear: trousers, jeans, skirt, socks, shorts, dress, shirt, T-shirt, jacket, sweater, shoes, trainers
things you wear when it's hot: shorts, T-shirt
things you wear when it's cold: socks, jacket, sweater

3 Aim: to present the words in the vocabulary box; to present the difference between the present simple and the present continuous.

- These vocabulary items are required for the questionnaire at the end of the lesson. This activity presents the concept question *What do you wear/are you wearing?* which allows the difference between the present simple and continuous to be illustrated.

4 Aim: to present the words in the vocabulary box.

- Ask the students to describe what the people in the photos are doing. Make sure they use the present continuous, which you often use to describe a scene, because it suggests that what is happening is temporary.

5 Aim: to practise listening; to present the difference between the present simple and continuous.

- Ask the students to listen and say who the people are. The language used on the tape identifies the four people by what they're doing and what they're wearing.
- Play the tape.

Answers
a – Harriet
b – John
c – Louise
d – Edward

6 Aim: to practise speaking; to provide an opportunity for a second listening.

- Ask the students to fill out the chart.
- Play the tape again for students to check.

	Harriet	John	Edward	Louise
What's he/she wearing?	jeans and a T-shirt	shirt and yellow tie	blue shirt, black trousers and trainers	black dress
What's she doing?	standing by the door talking to a middle-aged man	sitting in an armchair by the window	standing by the window and laughing	standing by the television and smiling

FUNCTIONS AND GRAMMAR

1 Aim: to focus on the uses of the present simple and continuous tenses.

● Ask the students to read the information in the box and then to do the exercises.

● Do this activity orally with the whole class.

Answers
1 wears
2 is smiling
3 is standing
4 is talking
5 are you working
6 stand up
7 smokes, isn't smoking
8 wears

2 Aim: to practise describing people.

● Ask the students to work in pairs and to describe people in the classroom.

3 Aim: to practise using the present continuous.

● Ask the students to write full answers to the questions about the people in the photo.

Possible answers
Harriet is wearing jeans and a T-shirt. She's standing by the door.
John is wearing a shirt and tie. He's sitting in the armchair.
Edward is wearing trainers, with a blue shirt and black trousers. He's holding something and smiling.
Louise is wearing a black dress. She's standing by the television and smiling.

4 Aim: to practise describing people.

● Ask the students to work in pairs and to describe the photos in the Communication activity. Explain that the two photos are similar but not the same.

READING AND SPEAKING

1 Aim: to practise reading for main ideas.

● The questionnaire focuses on clothes. Use it as a stimulus for discussion about what people wear and when. There will be minor differences of opinion between the students, so use this to explore attitudes about clothes. This will also have a contribution towards the syllabus of cross-cultural awareness in that if students are aware of differences within their own culture, they will be more aware of differences in other cultures.

2 Aim: to practise speaking.

● Ask the students to discuss their answers in pairs.

● Ask the students to discuss their answers with the whole class.

3 Aim: to practise reading.

● Ask the students to turn to the relevant Communication activity and read the analysis.

FUNCTIONS AND GRAMMAR

Describing people (2)

She's wearing a black dress. *She's sitting down.*
She's smiling.
He's the man sitting down in the armchair.

Present continuous or present simple

You use the present continuous to say what is happening now or around now.

*She's **wearing** a skirt.* *He's **living** in Oxford at the moment.*

You use the present simple to describe something which is true for a long time.

*She usually **wears** jeans.* *He **lives** in London.*

1 Choose the correct verb form.

1 He usually *wears/is wearing* a suit.
2 She *smiles/is smiling* at him.
3 Guy *is standing/stands* by the door.
4 Maria is the person who *is talking/talks* to Ken.
5 Where *are you working/do you work* at the moment?
6 You usually *stand up/are standing up* when you meet someone.
7 Ricardo *smokes/is smoking*, but he *doesn't smoke/isn't smoking* at the moment.
8 Pierre *wears/is wearing* brown shoes every day.

2 Work in pairs.

Student A: Choose someone in the class and describe him/her. Don't say who he/she is.

Student B: Listen to Student A's description of someone in the class. Can you guess who it is?

He's wearing jeans.
He's sitting down.

3 Write full answers to the questions in the chart in *Vocabulary and listening* activity 6.

Harriet's wearing jeans. She's standing by the door.

4 Work in pairs.

Student A: Turn to Communication activity 7 on page 100.

Student B: Turn to Communication activity 27 on page 105.

READING AND SPEAKING

1 Read the questionnaire and answer the questions.

What do your clothes say about you?

1 You see someone with blue hair wearing a yellow jacket and red trousers. What do you do?
a smile b laugh c wear the same clothes

2 You are going to an interview. What do you wear?
a jeans b a suit c something comfortable

3 You're going to work. What do you wear?
a trousers b trainers c a jacket

4 You're going to a party. What do you wear?
a a jacket b a T-shirt c a suit/dress

5 You're buying a new jacket. What colour do you buy?
a black b red c orange

6 You're buying clothes for cold weather. Which is more important?
a comfort b warmth c fashion

7 You want to give a good impression. Which style do you choose?
a comfortable but smart b smart and formal
c casual

8 What kind of clothes do you prefer?
a cheap b expensive c fashionable

9 You're going to play tennis with a friend. What do you do wear for the game?
a a tie b shorts
c a sweater

10 It's very hot at work or school. What do you do?
a wear shorts b take off your jacket or sweater
c do nothing

2 Work in pairs and talk about your answers to the questionnaire.

3 Turn to Communication activity 25 on page 105 and find out what your clothes say about you.

24 *I'm going to save money*

Going to; because **and** ***so***

READING AND LISTENING

1 Read the passage *My New Year's resolution*. Who do you think you can see in the photos?

> ## My New Year's resolution ...
>
> **1** 'I'm going to see my friends more often.' *Phil*
> **2** 'I'm going to save money.' *Harriet*
> **3** 'I'm going to change my job.' *Pete*
> **4** 'We're going to travel around Europe.' *Andrew and Mary*
> **5** 'We're going to have French lessons.' *Jill and Steve*
> **6** 'I'm going to spend more time with my parents.' *Jenny*
> **7** 'We're going to invite more friends for dinner.' *Henry and Celia*
> **8** 'I'm going to get fit.' *Kate*
> **9** 'I'm not going to take work home.' *Dave*
> **10** 'We're going to move.' *Judy and Frank*

2 Match the resolutions in 1 with the reasons below.

a We don't speak any foreign languages.
b We don't entertain very much.
c I hate my work.
d I stay at home all the time.
e We always stay in Britain for my holiday.
f Our house is too small.
g I never see my family.
h I want to spend more time with my children.
i I spend too much.
j I don't take enough exercise.

3 Listen to four people talking about their resolutions and the reasons. Find out who's speaking. Did you guess correctly in 2?

4 Work in pairs. Can you remember any other details about what the speakers said?

Now listen again and check.

24

GENERAL COMMENTS

New Year's resolution

You may need to explain that in Britain and the USA, it's traditional to make a New Year's Resolution which is a firm decision on 1 January to change one's life and to do things differently. It comes from a superstition that the New Year must begin happily and after a clean break with the past. It's equally traditional to break these resolutions a few days later.

You might like to tell the students about other superstitions concerning the New Year in Britain. In Northern England and Scotland it is believed that nothing must be removed from the house and people refuse to lend anything to neighbours until 2 January. In Scotland, if the first person who enters a home is tall and dark, he is believed to bring good luck. In Yorkshire and Lincolnshire you bring good luck if you have fair hair.

READING AND LISTENING

1 Aim: to practise reading for main ideas; to present examples of *going to*.

- Ask the students to read the resolutions and guess who they see in the photos.

Answer
Dave, Andrew and Mary

2 Aim: to practise reading for main ideas; to prepare the context for *because* and *so*.

- Ask the students to match the resolutions and the reasons.

Answers
1d 2i 3c 4e 5a 6g 7b 8j 9h 10f

3 Aim: to practise listening for main ideas.

- Ask the students to listen to four people and to decide who is speaking. Match the number of the speaker and their name.
- Play the tape.

Answers
1 Henry and Celia
2 Andrew
3 Kate
4 Judy and Frank

4 Aim: to practise speaking; to provide an opportunity for a second listening.

- Ask the students to discuss what they heard in detail.
- Ask several students to tell the whole class what the speakers said.
- Play the tape again and ask the students to listen and check.

Answers
Henry and Celia – their jobs take up a lot of time and they're tired at the weekend.
Andrew – will take a year off before university.
Kate – always drives to work and is going to start running.
Judy and Frank – have got a family and need more space.

GRAMMAR

1 Aim: to focus on the difference between the present simple and *going to*.

- Ask the students to read the information in the grammar box and then to do the exercises.
- You may like to do this activity with the whole class.

Answer
I go to work at nine o'clock. – something you do every day
I'm going to go to work at nine o'clock. – a future intention

2 Aim: to practise using *going to*.

- Ask the students to check their answers to *Reading and listening* activity 3 in pairs.
- Now ask the students to look at the passage and say what the people are going/not going to do.
- You may like to check this orally with the whole class.

3 Aim: to practise using *going to*.

- Tell the students about your plans for the weekend. (They don't have to be true.)
- Ask the students to ask and say what they're going to do this weekend.

4 Aim: to present the use of *because*.

● Remind the students that *because* is used before the cause of something.

● Ask the students to join some of the sentences in *Reading and listening* activity 1 and 2 with *because*.

Answers

1 Phil is going to see his friends more often **because** he stays at home all the time.
2 Harriet is going to save money **because** she spends too much.
3 Pete is going to change his job **because** he hates his work.
4 Andrew and Mary are going to travel around Europe **because** they always stay in Britain for their holiday.
5 Jill and Steve are going to have French lessons **because** they don't speak any foreign languages.
6 Jenny is going to spend more time with her parents **because** she never sees her family.
7 Henry and Celia are going to invite more friends for dinner **because** they don't entertain very much.
8 Kate is going to get fit **because** she doesn't take much exercise.
9 Dave is not going to take work home **because** he wants to spend more time with his children.
10 Judy and Frank are going to move **because** their house is too small.

5 Aim: to present the use of *so*.

● Remind the students that *so* is used before the consequence of something.

● Ask the students to join some of the sentences in *Reading and listening* activities 1 and 2 with *so*.

Answers

1 Phil stays at home all the time **so** he's going to see his friends more often.
2 Harriet spends too much **so** she's going to save money.
3 Pete hates his work **so** he's going to change his job.
4 Andrew and Mary always stay in Britain for their holiday **so** they're going to travel around Europe.
5 Jill and Steve don't speak any foreign languages **so** they're going to have French lessons.
6 Jenny never sees her family **so** she's going to spend more time with her parents.
7 Henry and Celia don't entertain very much **so** they're going to invite more friends for dinner.
8 Kate doesn't take much exercise **so** she's going to get fit.
9 Dave wants to spend more time with his children **so** he doesn't want to take work home.
10 Judy and Frank's house is too small **so** they're going to move.

VOCABULARY AND WRITING

1 Aim: to present the words in the vocabulary box.

● Check the students understand what the words mean.

● Do this collocation work with the whole class.

Answers

save money
take work home, take exercise
spend time
get fit
invite friends
change my job

2 Aim: to practise speaking.

● Ask the students to describe what they are going to do at the weekend using the verbs in the box.

3 Aim: to practise using *going to*.

● Ask five or six students what they're going to do before they finish *Reward* Elementary.

● Ask the students to think of two or three things they're going to do and write sentences.

4 Aim: to practise writing.

● Ask the students to think about why they are going to do the things they mentioned in 3 and write sentences.

5 Aim: to practise writing; to practise using *because*.

● Ask the students to join the sentences they have written with *because*.

6 Aim: to provide feedback for the students' resolutions.

● You may like to put some of these resolutions on the wall to remind people of their plans.

GRAMMAR

Going to

You use *going to* + infinitive:

– to talk about future intentions or plans.

I'm going to *see my friends more often.*

I'm not going to *take work home.*

– to talk about something which we can see now is sure to happen in the future.

I'm going to *have a baby.*

Because and *so*

You can join two sentences with *because* to describe a reason.

Judy and Frank are going to move ***because*** *their house is too small.*

You can join the same two sentences with *so* to describe a consequence.

Judy and Frank's house is too small ***so*** *they're going to move.*

1 Look at these sentences and explain the difference between them.

I go to work at nine o'clock.
I'm going to go to work at nine o'clock.

2 Work in pairs. Check your answers to *Reading and listening* activity 3. Say what the speakers are going to/not going to do.

3 Work in pairs. Say what you're going to do this weekend. Here are some ideas:

get up late   do some housework   play football
have a meal out   read the newspapers
meet some friends   watch television   go for a walk

4 Choose three sentences in *Reading and listening* activities 1 and 2 and join them with *because*.

Harriet is going to save money because she spends too much.

5 Choose three more sentences in *Reading and listening* activities 1 and 2 and join them with *so*.

Kate doesn't take enough exercise, so she's going to get fit.

VOCABULARY AND WRITING

1 Here are some verbs from this lesson. Can you remember which nouns or noun-phrases they went with in *My New Year's resolution*?

save   take   spend   get   invite   change

save money ...

2 Which phrases in 1 can you use to describe what you're going to do this weekend?

3 Write what you're going to do before you finish *Reward* Elementary.

I'm going to save more money.

4 Write why you're going to do the things you wrote in 3.

I want to go on holiday.

5 Join the sentences you wrote in 3 with the sentences you wrote in 4 using *because*.

I'm going to save more money because I want to go on holiday.

6 Make a class collection of resolutions and keep them safe. When you finish *Reward* Elementary, read them out and find out if anyone has kept their resolutions.

25 *Eating out*

***Would like*; talking about prices**

VOCABULARY AND LISTENING

1 Look at the words in the box. Which pictures of food and drink can you see on the menu?

burger Coke French fries ice cream sandwich salad pizza apple pie cheesecake kebab risotto pasta juice steak chocolate mousse coffee mayonnaise strawberry

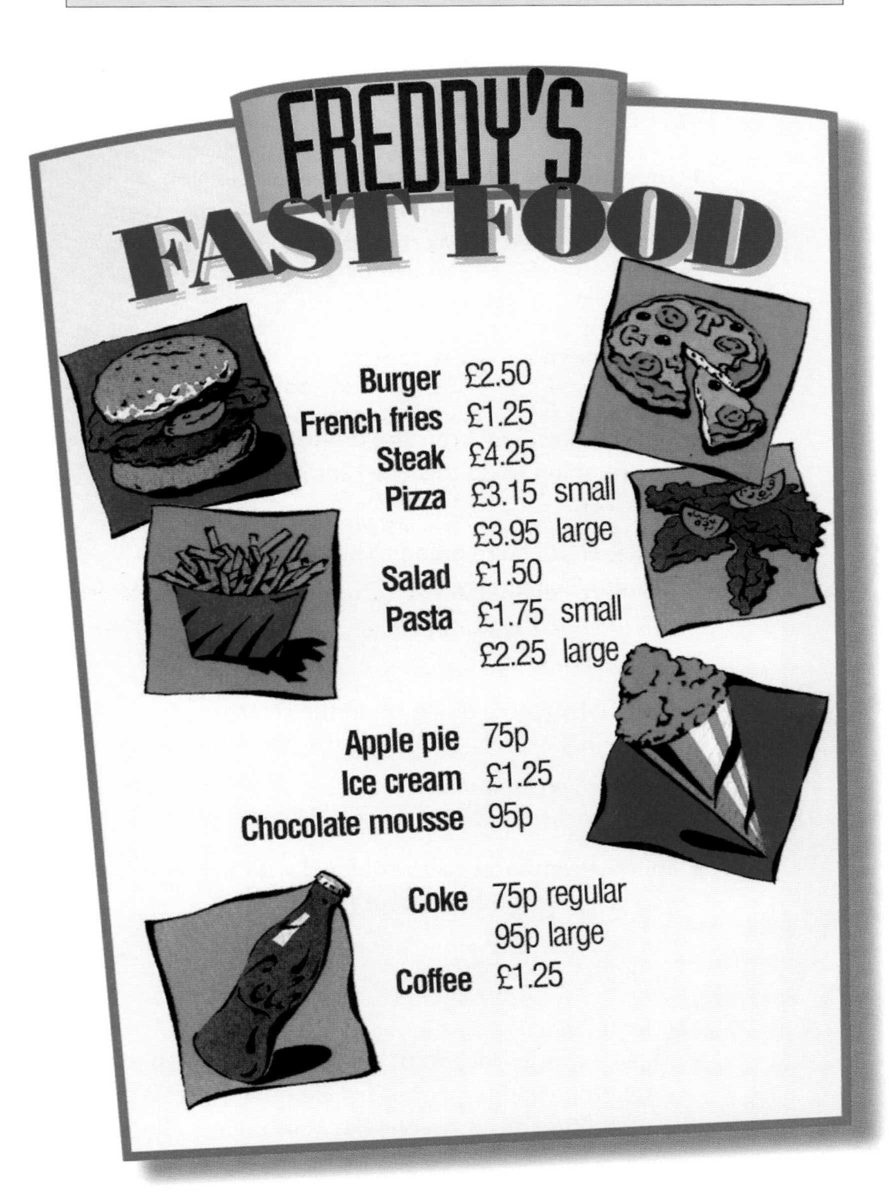

FREDDY'S
FAST FOOD
Burger £2.50
French fries £1.25
Steak £4.25
Pizza £3.15 small
£3.95 large
Salad £1.50
Pasta £1.75 small
£2.25 large
Apple pie 75p
Ice cream £1.25
Chocolate mousse 95p
Coke 75p regular
95p large
Coffee £1.25

2 Which is the odd word out?

1 burger Coke juice coffee
2 apple pie cheesecake ice cream pasta
3 pizza pasta steak risotto
4 French fries chocolate mousse salad mayonnaise

3 Look at the food items in the box. Say what you like.

4 Look at this conversation in a *fast food* restaurant. Decide where the sentences a – f go in the conversation.

ASSISTANT Good afternoon. Can I help you?
CUSTOMER (1) _____
ASSISTANT Would you like a regular or a large Coke?
CUSTOMER (2) _____.
ASSISTANT Would you like anything else?
CUSTOMER (3) _____
ASSISTANT What flavour would you like?
CUSTOMER (4) _____
ASSISTANT OK.
CUSTOMER (5) _____
ASSISTANT That's four pounds fifty.
CUSTOMER (6) _____
ASSISTANT Thank you.

a A regular, please.
b Good afternoon. Yes, I'd like a burger with fries and a Coke, please.
c Strawberry, please.
d How much is that?
e Yes, I'd like an ice cream, please.
f Here you are.

5 Listen and check your answers to 4.

25

GENERAL COMMENTS

American English

If your students found the American English words interesting in Lesson 23, here are some items to do with food.

British English	American English
aubergine	eggplant
biscuit	cookie
chips	French fries
crisps	chips
chocolate/sweets	candy
courgette	zucchini
green pepper	bell pepper

Cross-cultural awareness

The information about eating out in the USA is provided not necessarily because the students should know about restaurants there, but because it allows them to reflect on eating habits and customs in their own cultures.

VOCABULARY AND LISTENING

1 Aim: to present the words in the vocabulary box.

- Most of these words are likely to be used in different kinds of fast food restaurants. Ask your students if they prefer fast food or more traditional restaurants.
- Ask the students which items of food and drink they can see on the menu.

Answers

burger, French fries, ice cream, pizza, salad, coke

2 Aim: to practise using the words in the vocabulary box.

- Ask the students to choose the odd word out.

Answer

1 burger
2 pasta
3 steak
4 chocolate mousse

- Ask the students to make up their own odd words out game and to write the words on pieces of paper. When they're ready, they should swap their papers with another student, who should then find the odd word out.

3 Aim: to practise using the new vocabulary.

- Ask the students which items of food and drink they like.

4 Aim: to prepare for listening.

- This is a fairly predictable conversation to be heard in a fast food outlet. Ask the students to read the conversation and decide where the sentences should go.

Answers

1 b 2 a 3 e 4 c 5 d 6 f

5 Aim: to practise listening for specific information.

- Play the conversation. Ask the students to listen and check their answers to 4.

FUNCTIONS

1 Aim: to focus on the difference between *like* and *would like.*

● Ask the students to read the information in the functions box and then to do the exercises.

● You can do this activity orally with the whole class.

Answers
1b 2a 3a 4b 5b 6a

2 Aim: to practise speaking.

● Ask the students to act out the conversation in pairs.

● When they're ready, ask three or four pairs to act out the conversation in front of the class.

3 Aim: to practise the language for talking about prices.

● Ask the students how much certain items cost in their country. This will provide the model the students will need for this activity.

● Ask the students to work in pairs and to ask and say how much things cost on the menu.

SOUNDS

1 Aim: to focus on the pronunciation of *like* and *would like.*

● 📼 Play the tape and ask the students to listen to the sentences and to tick the ones they hear.

Answers
1b 2a 3b 4a 5a 6a

● Ask the students to read the sentences aloud.

READING AND SPEAKING

1 Aim: to pre-teach difficult vocabulary.

● Make sure the students know the meaning of the underlined words.

2 Aim: to practise reading for specific ideas.

● Ask the students to read the passage to find out what it says about the points mentioned.

Answers
types of restaurants: fast food, coffee shop, diner, family restaurant, top class restaurant
where to sit: coffee shop - you sit at the counter or at a table; family restaurant - a waitress shows you where to sit; top class restaurant - a waiter shows you where to sit
who to pay: fast food restaurant - you pay the person who serves you; coffee shop - you pay the cashier; top class restaurant - you pay the waiter
how much to tip: fast food restaurant - no need to tip; family restaurant - add 15 per cent to the bill; top class restaurant - add 15 per cent to the bill
other advice: fast food restaurant - put the empty container and paper in a rubbish bin; coffee shop - you don't wait for the waitress to show you where to sit; family restaurant - you don't need to tell the waitress your name, if you don't eat everything, the waitress will bring a doggy bag; top class restaurant - you can only refuse the wine you taste if it is bad.

3 Aim: to check answers to 2; to practise speaking; to practise cross-cultural comparison.

● Ask the students to work in pairs to check their answers.

● Lead a discussion with the whole class about eating out in the USA and the students' countries.

FUNCTIONS

Would like

You use *would like* + infinitive to ask for something politely.

I'd like *a burger and a Coke, please.*
Would you like *a regular or a large Coke?*
I'd like *an ice cream, please.*
What flavour ***would you like?***

Remember

– you use *like* to say what you like all the time.
I ***like*** *Coke.* (= always)

– you use *would like* to say what you want now.
I'd ***like*** *a Coke.* (= now)

Talking about prices

How much is it? *Two pounds ninety-nine.*

1 Choose the correct answer.

1 Would you like a drink?
a No, I don't like a drink. b No, thank you.
2 What would you like to eat?
a I'd like a pizza. b I like pizza.
3 Would you like some coffee?
a Yes, please. b Yes, I like some wine.
4 Do you like pasta?
a Yes, please. b Yes.
5 Can I help you?
a Yes. b Yes, I'd like a burger.
6 Would you like to order?
a Yes, a burger, please. b Yes, I like burger.

2 Work in pairs and act out the conversation in *Vocabulary and listening* activity 4. Choose other items from the menu.

3 Work in pairs. Look at the menu and ask and say how much things cost.

SOUNDS

Listen and tick (✓) the sentences you hear.

	a	b
1	I like beer.	I'd like a beer.
2	He likes ice cream.	He'd like ice cream.
3	We like the bill.	We'd like the bill.
4	I like pizza.	I'd like pizza.
5	She'd like a salad.	She likes a salad.
6	Would you like pasta?	Do you like pasta?

READING AND SPEAKING

1 You're going to read a passage about eating out in the USA. First, check you know what the underlined words in the passage mean.

2 Read *Eating out in the USA* and find out what it says about:

– types of restaurants – where to sit – who to pay
– how much to tip – other advice

EATING OUT IN THE USA

In the USA, there are many types of restaurant. Fast food restaurants are very famous, with McDonalds and Kentucky Fried Chicken in many countries around the world. You look at a menu above the counter, and say what you'd like to eat. You pay the person who serves you. You take your food and sit down or take it away. When you finish your meal, you put the empty container and paper in the rubbish bin. There's no need to leave a tip.

In a coffee shop you sit at the counter or at a table. You don't wait for the waitress to show you where to sit. She usually brings you coffee when you sit down. You tell her what you'd like to eat and she brings it to you. You pay the cashier as you leave. A diner is like a coffee shop but usually looks like a railway carriage.

In a family restaurant the atmosphere is casual, but the waitress shows you where to sit. Often the waitress tells you her name, but you don't need to tell her yours. If you don't eat everything, your waitress gives you a doggy bag to take your food home. You add an extra fifteen per cent to the bill as a tip.

In top class restaurants, you need a reservation and you need to arrive on time. The waiter shows you where to sit. If you have wine, he may ask you to taste it. You can only refuse it if it tastes bad, not if you don't like it. When you get your bill, check it and then add fifteen to twenty per cent to it as a tip. You pay the waiter.

3 Work in pairs and check your answers. Compare eating out in the USA with eating out in your country.

Progress check 21–25

VOCABULARY

1 When you see a word you don't understand, stop and ask yourself these questions:

– what is its part of speech?
– can I guess what it means?
– can the rest of the passage help me understand it?

Try not to use a dictionary or ask your teacher every time.

Look at this extract from the passage in Lesson 25. Some of the difficult words are missing. Think about your answers to the questions above. Don't try to remember the exact word.

> In the USA, there are several types of restaurant. *Fast food* restaurants are very famous, with McDonalds and Kentucky Fried Chicken in many countries around the world. You look at a _____ above the _____, and say what you'd like to eat. You pay the person who serves you. You take your food and sit down or take it away. When you finish your meal, you put the _____ container and paper in the rubbish bin. There's no need to leave a tip.
>
> In a *coffee shop* you sit at the counter or at a table. You don't wait for the _____ to show you where to sit. She usually brings you coffee as soon as you sit down. You pay the _____ as you leave. A *diner* is like a coffee shop but usually looks like a railway carriage.

Now look back at page 59 to find out what the missing words are.

2 Find ten words in the word puzzle. They go in two directions (↓) and (→). Five words are to do with food and drink and five words are to do with clothes.

c	s	o	c	k	a	s	c	v	g
h	a	m	b	u	r	g	e	r	a
o	s	d	f	g	t	k	l	m	j
c	h	e	e	s	e	c	a	k	e
o	d	f	v	a	e	z	p	s	a
l	c	x	e	l	s	f	p	h	n
a	f	w	q	a	h	v	l	i	s
t	r	y	u	d	o	n	e	r	e
e	t	h	h	t	e	m	u	t	q
t	i	g	h	t	s	k	r	h	x

3 Choose one of the other topics in Lessons 21 to 25 and make a word puzzle.

When you're ready, work in pairs and do each other's word puzzles.

GRAMMAR

1 Complete the sentences with *ago, from, to, last, yesterday, during, at.*

1 He started work _____ morning.
2 We went camping _____ the summer.
3 I started learning French _____ year.
4 He rang me five minutes _____.
5 It was open _____ nine _____ five.
6 He left _____ the end of the week.

Progress check 21 - 25

GENERAL COMMENTS

You can work through this Progress check in the order shown, or concentrate on areas which have caused difficulty in Lessons 21 - 25. You can also let the students choose the activities they would like or feel the need to do.

VOCABULARY

1 Aim: to present some techniques for dealing with unfamiliar words.

- Ask the students to read the advice on dealing with unfamiliar words. Explain to them (again, if necessary) that they have to develop these techniques because they won't always have a dictionary or a teacher nearby to explain everything they don't understand. Furthermore, if they do look up every word, their reading speed is greatly reduced.

- This activity is designed to show the students that they don't always have to know what a word means to understand the general sense of the passage. In this case, they aren't even able to see the whole passage.

Answers
menu, counter, empty, waitress, bill

2 Aim: to revise words for food, drink and clothes.

- Ask the students to do this on their own, and then ask them to check their answers.

Answers
Food and drink: chocolate, salad, apple, hamburger, cheesecake
Clothes: shoes, shirt, jeans, sock, tights

- Find out how many students found all ten words.

3 Aim: to revise vocabulary from Lessons 21 to 25.

- Ask the students to make their own word puzzles.

- When they're ready, ask the students to do each other's word puzzles.

GRAMMAR

1 Aim: to revise expressions of time.

Answers
1 He started work yesterday morning.
2 We went camping during the summer.
3 I started learning French last year.
4 He rang me five minutes ago.
5 It was open from nine to five.
6 He left at the end of the week.

2 **Aim: to revise the difference in meaning between the present simple and continuous.**

Answers
1 wear 2 is standing 3 smokes
4 shake 5 is smiling 6 speak

3 **Aim: to revise the use of *going to.***

- Ask the students to write five plans or intentions.

4 **Aim: to revise the use of *because.***

Answers
1 I'd like something to eat **because** I'm hungry.
2 Would you like pasta **because** we haven't got any pizza?
3 I'm going to see Tom Cruise's new film **because** I like him.
4 I'm going to live in London **because** I like the people there.
5 He can't see **because** he hasn't got his glasses.
6 We're not going away this year **because** we went on holiday last year.

5 **Aim: to revise the use of *so.***

Answers
1 I'm hungry **so** I'd like something to eat.
2 We haven't got any pizza **so** would you like pasta?
3 I like Tom Cruise **so** I'm going to see his new film.
4 I like the people in London **so** I'm going to live there.
5 He hasn't got his glasses **so** he can't see.
6 We went on holiday last year **so** we're not going away this year.

SOUNDS

1 **Aim: to focus on homophones.**

- Play the tape. Explain that words which sound the same but have a different spelling are called homophones.

Answers
meet – meat, their – there, know – no, sea – see, son – sun, write – right, our – hour, buy – by, eye – I, for – four, knows – nose, too – two

2 **Aim: to focus on /uː/ and /ʊ/ .**

- Ask the students to say the words aloud.
- Ask the students to put the words in two columns.
- Play the tape for the students to check. Ask them to repeat the words aloud.

Answers
/uː/: food, shoe, you, soup, do, cool, juice, boot
/ʊ/: good, cook, book, foot, put

- You may find that people from certain regions of Britain will pronounce these words differently.

3 **Aim: to focus on stressed syllables in words.**

- Ask the students to underline the stressed syllables in the words.

Answers
banana, cabbage, potato, bacon, trousers, trainers, cashier, risotto, salad, hamburger, toothpaste

- Play the tape. Ask the students to check their answers and say the words aloud as they hear them.

4 **Aim: to focus on polite intonation.**

- Play the tape and ask the students to listen and say which speaker sounds polite.

Answers
Polite:
1 Can I help you?
3 Would you like a drink?

WRITING AND SPEAKING

1 **Aim: to practise using the past simple; to practise writing.**

- Ask the students to write questions about well-known people, places or historical facts.
- You may like to ask them to do this for homework.

2 **Aim: to practise speaking.**

- Ask the students to do their quizzes in groups of four.
- Find out which group gets the highest score.

2 Choose the correct verb form.

1 I *wear/am wearing* jeans most of the time.
2 He's the man who *stands/is standing* next to Jim.
3 He *smokes/is smoking* twenty cigarettes a day.
4 I always *shake/am shaking* hands when I meet someone.
5 At the moment she *smiles/is smiling* at him.
6 I *speak/am speaking* fluent English.

1 I wear jeans most of the time.

3 Write five things which you're going to do next month.

I'm going to visit my sister next month.

4 Join the two sentences by rewriting them with *because.*

1 I'm hungry. I'd like something to eat.
2 I'm afraid we haven't got any pizza. Would you like pasta?
3 I like him. I'm going to see Tom Cruise's new film.
4 I'm going to live in London. I like the people there.
5 He can't see. He hasn't got his glasses.
6 We went on holiday last year. We're not going away this year.

1 I'd like something to eat because I'm hungry.

5 Rewrite the sentences in 4 with *so.*

1 I'm hungry so I'd like something to eat.

SOUNDS

1 Listen and repeat the following words.

meet their know sea son write our buy eye right no sun for there by I four knows too hour see meat nose two

Write the pairs of words which sound the same but have different spelling.

2 Say these words aloud. Is the underlined sound /ʊ/ or /uː/?

food shoe you good soup cook book do foot cool put juice boot

Now listen and check. As you listen, say the words aloud.

3 Look at these words. Underline the stressed syllable.

banana cabbage potato bacon trousers trainers cashier risotto salad hamburger toothpaste

Listen and check. As you listen, say the words aloud.

4 Listen to these questions. Put a tick (✓) if you think the speaker sounds polite.

1 Can I help you?
2 What would you like to eat?
3 Would you like a drink?
4 Would you like anything else?

Now say the questions aloud. Try to sound polite.

WRITING AND SPEAKING

1 Work in groups of three or four. You're going to prepare a quiz about important historical facts.
Write at least ten questions about important events/people in history. Think about:

– life and death of famous people
– inventions and discoveries
– wars
– political events
– artistic creations

2 Work with another group. In turn, ask and answer questions from your quizzes. You score one point for each correct answer. The group with the most points is the winner.

26 Can I help you?

Reflexive pronouns; saying what you want to buy; giving opinions; making decisions

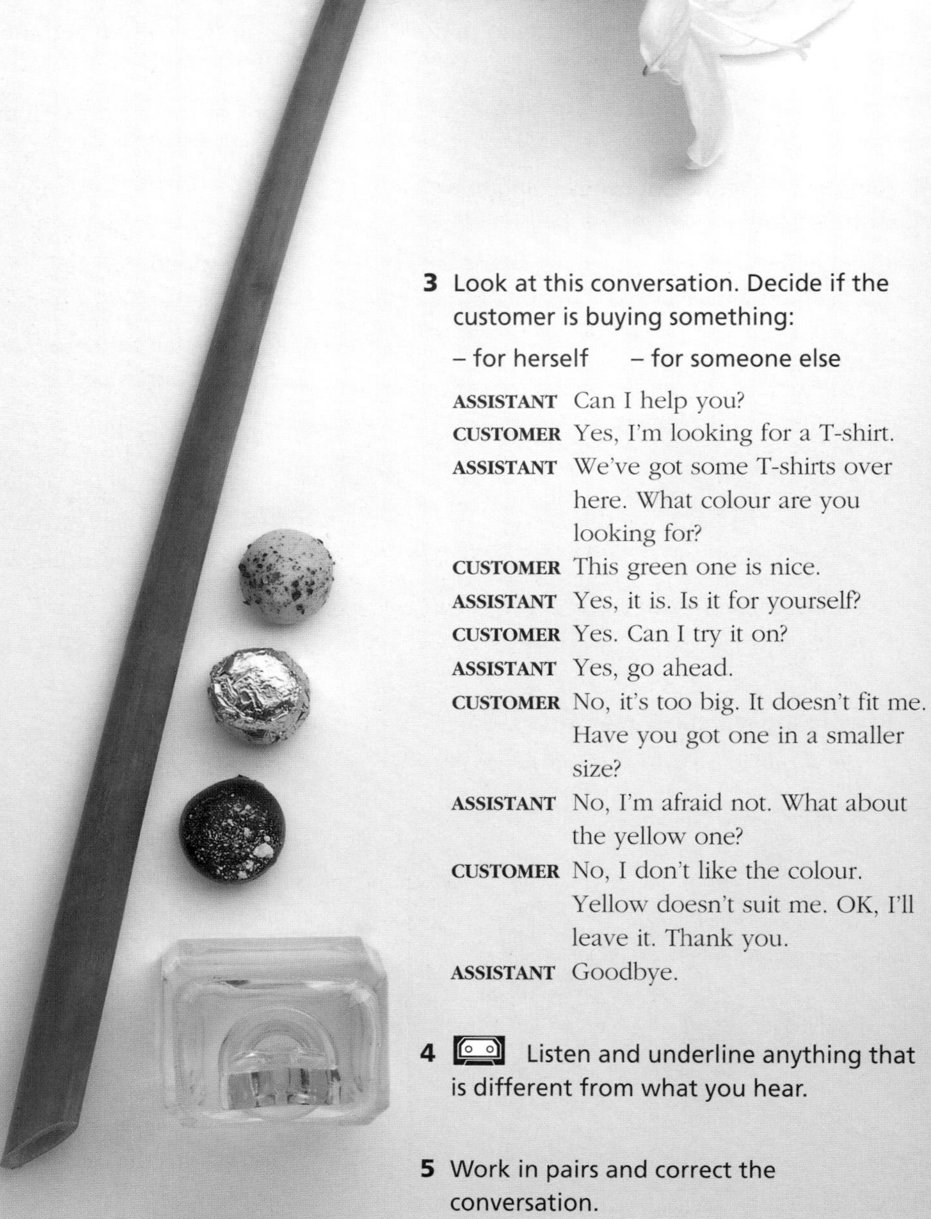

SPEAKING AND LISTENING

1 Match the sentences 1 – 4 with the sentences a – d to make four conversations.

1 Can I carry that for you?
2 Did you make it?
3 Are you buying this for yourself?
4 They've got themselves a new one.

a No, it's for a friend.
b Have they? What kind?
c No, Geoff made it himself.
d No, it's OK. I can carry it myself.

2 Work in pairs and check your answers to 1. Choose one conversation and write two or three sentences before and after it. Act out your conversation to the rest of the class.

3 Look at this conversation. Decide if the customer is buying something:

– for herself – for someone else

ASSISTANT Can I help you?
CUSTOMER Yes, I'm looking for a T-shirt.
ASSISTANT We've got some T-shirts over here. What colour are you looking for?
CUSTOMER This green one is nice.
ASSISTANT Yes, it is. Is it for yourself?
CUSTOMER Yes. Can I try it on?
ASSISTANT Yes, go ahead.
CUSTOMER No, it's too big. It doesn't fit me. Have you got one in a smaller size?
ASSISTANT No, I'm afraid not. What about the yellow one?
CUSTOMER No, I don't like the colour. Yellow doesn't suit me. OK, I'll leave it. Thank you.
ASSISTANT Goodbye.

4 Listen and underline anything that is different from what you hear.

5 Work in pairs and correct the conversation.

26

GENERAL COMMENTS

Customer service relationships

You may like to know and to share with your students that in Britain and America, the relationship between customers and shop assistants is friendly and polite, and dealings are often conducted on a basis of equality rather than servility or indifference. All the verbal exchanges will contain the usual, sometimes elaborate, politeness formulae, and will avoid abruptness and direct instructions.

When you enter a shop which has assistants (unlike supermarkets), you will very often be greeted with *Can I help you?* If you think the assistant can help you, then say so, but you must feel under no obligation to buy anything. If you prefer not to have help, you can say *It's all right, I'm just looking*. At this point, the assistant will leave you alone. In most shops you are usually welcome to pick up goods and examine them before you buy them and without asking for permission, as long as they are within reach. Goods behind the counter are there in order to be protected against theft or damage.

Find out if the students think customer service relationships are similar in their country.

SPEAKING AND LISTENING

1 Aim: to practise speaking; to present reflexive pronouns.

- These exchanges can generate a certain amount of discussion work in the process of matching the two parts, and can act as a stimulus for an improvised conversation.
- You may like to do this activity with each part of the conversation written on a separate piece of paper. Students should learn their part and go round the class saying it until they have found a suitable partner.

Answers
1d 2c 3a 4b

- Correct this activity with the whole class.

2 Aim: to practise speaking.

- Ask the students to extend their conversations with two or three sentences before and after their exchange.
- When they are ready, ask the students to act out their conversations in front of the class.
- You may like to ask your students to do this activity for homework.

3 Aim: to prepare for listening.

- Ask the students to read the conversation and to decide if the customer is buying something for herself or for someone else.

Answer
For herself

- You may like to answer a few questions about difficult vocabulary.

4 Aim: to practise listening for specific information; to present the language of saying what you'd like, making decisions and giving opinions.

- Play the tape and ask the students to underline anything which is different from what they hear.

Answers

ASSISTANT	Can I help you?
CUSTOMER	Yes, I'm looking for a T-shirt.
ASSISTANT	We've got some T-shirts over here. What colour are you looking for?
CUSTOMER	This green one is nice.
ASSISTANT	Yes, it is. Is it for yourself?
CUSTOMER	Yes, it is. Can I try it on?
ASSISTANT	Yes, go ahead.
CUSTOMER	No, it's too big. It doesn't fit me. Have you got one in a smaller size?
ASSISTANT	No, I'm afraid not. What about the yellow one?
CUSTOMER	No, I don't like the colour. Yellow doesn't suit me. OK, I'll leave it. Thank you.
ASSISTANT	Goodbye.

5 Aim: to focus on the target language; to practise speaking.

- Ask the students to correct the conversation so it follows what they heard. You may need to play the tape again.

GRAMMAR AND FUNCTIONS

1 Aim: to focus on reflexive pronouns.

- Ask the students to read the information in the grammar and functions box and then to do the exercises.
- Ask the students if their language has reflexive pronouns or has some other way of expressing the same idea.
- Ask the students to complete the sentences.

Answers

1 myself	2 ourselves	3 herself
4 yourself	5 themselves	6 himself

2 Aim: to practise speaking; to practise using the target language.

- Ask the students to act out the conversation in *Speaking and listening* activity 3.

VOCABULARY AND LISTENING

1 Aim: to present the words in the vocabulary box.

- Ask the students to work in pairs and to decide which items they can see in the photos.
- Check the answers with the whole class.

Answers
chocolates, flowers, perfume, cake

2 Aim: to present the words in the vocabulary box.

- Ask the students to match the different types of 'containers' with the items in 1.

Answers
a box of chocolates, a packet of biscuits, a bunch of flowers, a bottle of milk, a bottle of perfume, a pair of jeans, a bar of soap

3 Aim: to prepare for listening; to practise speaking.

- Ask the students to talk about which items they buy for themselves and which they buy as gifts for other people.

4 Aim: to practise listening for specific information.

- Play the tape and ask the students to decide what each person is buying.

Answers
Conversation 1: perfume
Conversation 2: chocolates

5 Aim: to check comprehension; to practise listening.

- Ask the students to work in pairs to check their answers, and to say who the speakers are buying the items for.
- Play the tape so the students can check again.

Answers
Conversation 1: for his wife
Conversation 2: for a friend

6 Aim: to practise speaking.

- Ask the students to act out a conversation based on items in the photos.
- When they are ready, ask the students in turn to perform their conversations to the rest of the class.

SOUNDS

1 Aim: to focus on polite intonation.

- Remind your students that customer service relationships are usually polite.
- Ask the students to listen to the conversations in *Vocabulary and sounds* activity 4 and to say in which conversation the customer sounds polite.

Answer
Conversation 2

SPEAKING

1 Aim: to practise speaking.

- Ask the students to work in groups and to discuss their answers to the questions.

2 Aim: to practise speaking.

- Ask the students to visit other groups and to find out their views on shopping.

3 Aim: to practise speaking.

- Ask the students to report back to their group and tell them what they've learnt.
- You may like to ask each group in turn to give feedback on their discussion to the whole class.

GRAMMAR AND FUNCTIONS

> **Reflexive pronouns**
> **You usually use a reflexive pronoun when the subject and the object of a sentence are the same.**
> *myself yourself himself herself itself ourselves yourselves themselves*
> ***I** can carry it **myself.***
>
> **Saying what you want to buy**
> *I'd like a ...*
> *I'm looking for ...*
> *Can I try it on?*
> *Have you got a/any ...?*
> *Have you got it in another colour?*
>
> **Giving opinions**
> *It's too big/small/long/short.*
> *It doesn't suit me.*
> *It doesn't fit me.*
> *I don't like the colour.*
>
> **Making decisions**
> *Can I try it in a different size?*
> *I'll have this/these. I'll take it/them.* *I'll leave it.*

1 Complete the sentences with a reflexive pronoun.

1 Was that T-shirt a gift? No, I bought it for _____.
2 Tim and I enjoyed _____ at the disco last night.
3 She doesn't live by _____ . She lives with friends.
4 Can I have some coffee? Yes, would you like to serve _____?
5 They taught _____ to speak Russian. They didn't have lessons.
6 He's unhealthy and smokes too much. He doesn't look after _____.

2 Work in pairs and act out the conversation in *Speaking and listening* activity 3.

VOCABULARY AND LISTENING

1 Look at the words in the box. Which items can you see in the photos?

chocolates biscuits cakes flowers milk perfume jeans soap

2 Which words in 1 do the words below go with?

box packet bottle bunch pair bar

a box of chocolates, ...

3 Work in pairs. Which items do you buy for yourself? Which items do you buy as gifts for other people?

I often buy chocolates for myself.

4 Listen to two conversations. Put the number of the conversation by the items in 1 which the people are buying.

5 Work in pairs and check your answers to 4. Who are they buying the items for?

Now listen again and check.

6 Work in pairs.

Student A: You're a shop assistant. You sell the items in the photos.
Student B: You're a customer. You want to buy something in the photos.

Act out a conversation. Use the conversations in 4 to help you.

SOUNDS

Listen to the conversations in *Vocabulary and listening* activity 4 again. In which conversation does the customer sound polite and friendly?

SPEAKING

1 Work in groups of three or four and discuss your answers to the questions.

1 Do you like shopping?
2 When you go shopping, do you usually go by yourself?
3 Do you know what you want to buy?
4 Do you usually buy things for yourself?
5 How often do you buy things for other people?
6 Who do you buy gifts for?
7 When do you give gifts?
8 What do you buy for gifts?

2 Find out what other students' answers are.

3 Report back to your group.

Mario often buys things for himself.

27 Whose bag is this?

Whose; possessive pronouns; describing objects

VOCABULARY AND SPEAKING

1 Complete the sentences below the pictures with words from the box.

short long square round small large rectangular heavy light

2 Say what the things in the pictures are made of. Use the words in the box.

glass leather metal plastic paper wood

The ruler is made of plastic.

3 Work in pairs. Choose an object in the classroom. In turn, ask and answer questions to try and guess the object your partner is describing.

A *It's square.*
B *Is it a book?*
A *No. It's made of glass.*
B *Is it the window?*
A *Yes.*

4 Match the conversations and the pictures below.

A Excuse me, is this yours?
B No, it isn't mine. It's his.

A Whose is this?
B It's theirs.

a *It's long.*
b *It's short.*
c *It's square.*
d *It's* ________.
e *It's* ________.
f *It's* ________.
g *It's* ________.
h *It's* ________.
i *It's* ________.

27

GENERAL COMMENTS

Expressions with *made*

You use *made of* to describe the material used to make something. *It's made of leather.* You use *made out of* to focus on the process of manufacture. *The child's doll was made out of sticks, wool, cloth and cotton wool.* When the material completely changes in the process of making something, you use *made from. Beer is made from hops, barley and water.* It's probably better to focus on *made of,* and to treat it as a prepositional phrase rather than a present simple passive (see Lesson 29).

VOCABULARY AND SPEAKING

1 Aim: to present the words in the vocabulary box.

● Draw shapes on the board and ask students to label them with words from the box.

● Ask the students to complete the sentences with words from the box.

Answers
d It's round e It's rectangular f It's light
g It's large h It's small i It's heavy

2 Aim: to present the words in the vocabulary box.

● Ask the students to say what the things in the drawings are made of.

Answers
The long ruler is made of wood.
The yellow square is made of paper.
The ball is made of leather.
The diary is made of leather.
The bag is made of plastic.
The glass is made of glass.
The weight is made of metal.

● Go round the classroom pointing at objects and asking *What's it made of?* Elicit a suitable response.

3 Aim: to practise using the new vocabulary.

● Ask the students to work in pairs. Student A should choose an object in the classroom, but not say what it is. Student B should ask questions, trying to guess what object Student A is thinking of.

4 Aim: to present possessive pronouns.

● Ask the students to match the conversations and the pictures.

GRAMMAR AND FUNCTIONS

1 Aim: to focus on the formation of possessive pronouns.

- Ask the students to read the information in the grammar and functions box, and then to do the exercises.
- Ask the students to say the possessive pronouns aloud.
- Ask the students to decide which letter is often added to the possessive adjective to form the possessive pronoun.

Answer
The letter is *s*. The exceptions are *mine* and *his.*

2 Aim: to focus on the difference between the possessive adjective and possessive pronoun.

- Ask the students to choose the correct word. You may like to do this orally with the whole class.

Answers

1 mine	2 my	3 her, mine
4 Who's	5 their, ours	6 Whose

3 Aim: to practise using the language for describing objects.

- You may like to play a round of this game with two or three students in front of the class.
- Ask students to carry on playing this game in pairs.

LISTENING AND SPEAKING

1 Aim: to prepare for listening.

- You may like to explain that a Lost Property office is the place where people might find possessions, which they've lost.
- Ask the students to match the numbers and the questions.

Answers
1e 2f 3h 4a 5d 6i 7b 8g 9c

2 Aim: to practise listening for specific information.

- Play the tape and ask the students to underline anything which is different from what they hear.

Answers

		Lost Property
1	Name	Ms Jill Fairfield
2	Address	32 Burn Road, Manchester
3	Telephone	678 5463
4	Lost article	bag
5	Date of loss	21 July
6	Time of loss	about 10 in the morning
7	Place of loss	Chester Market
8	Description	large, square, made of black nylon
9	Contents	a purse, a calculator, an address book a comb and a newspaper

3 Aim: to practise speaking.

- Ask the students to correct the information.
- Play the tape again.

4 Aim: to practise speaking.

- Ask the students to act out the conversation in pairs.
- Ask one or two pairs of students to act out the conversation in front of the whole class.

5 Aim: to practise listening for specific information.

- Ask the students to listen and put the number of the speaker by the answer to each question.
- Play the tape.

Answers

Conversation 1
Ken Hamilton, 13, Dock Lane, London,
75859, coat, 31 May, 4 in the afternoon,
on the train, red leather

Conversation 2
Mary Walter, 21, Tree Road, Leeds,
75889, bag, 20 March, 2 in the afternoon,
in the supermarket, black plastic, shopping, purse

Conversation 3
Ian Joseph, 33, James Street, Bath,
56778, box of cigars, 17 October,
11 in the morning, on the bus, wooden, square

6 Aim: to practise speaking.

- Ask the students to act out the three conversations, using the questions in 1 and answers in 5 as prompts.
- Ask the students to learn their lines for homework. They can then perform the conversations in front of the whole class in the next lesson.

GRAMMAR AND FUNCTIONS

Whose

You use *whose* to ask who something belongs to.

***Whose** bag is this?*
***Whose** shoes are these?*

Possessive pronouns

You use possessive pronouns to say who something belongs to.

mine yours his hers ours theirs
*Whose bag is this? It's **mine.***
*Whose shoes are these? They're **his.***

Describing objects

You don't usually put more than two or three adjectives together.

What's it like?
It's a small, plastic ruler.

1 Look at these possessive adjectives.

my your his her its our their

Which letter do you add to the possessive adjective to make a possessive pronoun? Which possessive pronouns are the exception?

2 Choose the correct word in these sentences.

1 Whose is this? It's *my/mine.*
2 Where did I put *my/mine* bag?
3 These aren't *her/hers* shoes. They're *my/mine.*
4 *Whose/who's* got my pen?
5 Are these *their/theirs* books? No, they're *our/ours.*
6 *Whose/who's* coat is this?

3 Work in groups of four or five. Put two personal possessions on a desk or in a bag. Go round, in turn, holding up or taking out a possession, asking and saying who it belongs to.

A Whose is this?
B It's mine. And whose is this?
C It's his.

LISTENING AND SPEAKING

1 You're going to hear a conversation in a Lost Property office. Look at the form below. Match the items 1 – 9 and the questions a – i.

Lost property		
1	Name	Mrs Joan Fairfield
2	Address	22, Burn Lane, Macclesfield
3	Telephone	678 5463
4	Lost article	bag
5	Date of loss	20 July
6	Time of loss	10am
7	Place of loss	Chester market
8	Description	small, square and it was made of black leather
9	Contents	a purse, a calculator, an address book, a comb

a What did you lose?
b Where did you lose it?
c What was in it?
d What date did you lose it?
e What's your name?
f What's your address?
g What's it like?
h What's your telephone number?
i What time did you lose it?

2 Look at the Lost Property form. Listen and underline any information which is different from what you hear.

3 Work in pairs. Correct any information which was different.

Listen again and check.

4 Work in pairs. Act out the conversation you heard in 2. Use the form and the questions to help you.

5 Listen to three more conversations in a Lost Property office. Put the number of the conversation by the answer to the questions below. There is one extra answer for each question.

John Smith	Ian Joseph	Mary Walter	Ken Hamilton
13, Dock Lane, London	21, Tree Road, Leeds	33, James Street, Bath	45, Old Road, Oxford
56778	56983	75859	75889
bag	wallet	coat	box of cigars
20 March	31 May	12 June	17 October
11am	2pm	4pm	11pm
train	bus	supermarket	chemist
rectangular, nylon	black, plastic	red leather	wooden, square

6 Work in pairs. Act out the conversations in 5. Change your partner for each conversation.

28 What's the matter?

Asking and saying how you feel; sympathising; *should, shouldn't*

VOCABULARY AND LISTENING

1 Look at the words in the box. Find:

- two types of medicine
- six complaints or illnesses
- five adjectives to describe how you feel
- seven parts of the body

arm aspirin back cold (noun)
cough cough medicine dizzy
faint finger foot hand
headache ill leg sick
sore throat stomach ache
temperature tired toe

2 Listen to three conversations. Say what's wrong with each person.

The first person has got a headache.

3 Put the number of the person by what you think he/she should do.

- ☐ go to bed
- ☐ stay at home
- ☐ drink plenty of water
- ☐ stop smoking
- ☐ get some exercise
- ☐ go to the doctor
- ☐ keep warm
- ☐ eat nothing for 24 hours
- ☐ lie down

FUNCTIONS AND GRAMMAR

Asking and saying how you feel

What's the matter?	*I don't feel very well.*
Are you all right?	*I feel sick.*
	I've got a headache.
	My back hurts.

Sympathising

Oh dear! *What a pity!* *Oh, I am sorry!*

Should, shouldn't

***Should* and *shouldn't* are modal verbs.**

You use *should* and *shouldn't* to give advice.

*You **should** go to bed.* *You **shouldn't** go to work.*

(For more information about modal verbs, see Grammar review page 111.)

1 Look at the functions and grammar box and answer the questions.

1. What follows *I feel* – an adjective or a noun?
2. What follows *I've got* – an adjective or a noun?
3. What comes before *hurts* a person or a part of the body?

2 Work in pairs and say what the people in the conversations in *Listening* activity 2 should or shouldn't do.

I think he should go to bed.
He shouldn't go to work.

3 Work in pairs.
Student A: Turn to Communication activity 16 on page 102.
Student B: Turn to Communication activity 21 on page 104.

READING AND WRITING

1 Read and answer the questions for your country.

1. When you're ill, do you go to a specialist who knows about your illness or your local doctor?
2. Are there both men and women doctors?
3. Where do you get medicine in your country?
4. Do you ever go to the doctor if you're well?
5. Do doctors visit you at home?
6. What do you do in an emergency?
7. Do friends and relatives visit you in hospital?
8. Do you pay for medical treatment?

2 Read the advice leaflet *The nation's health* and find the answers to the questions for Britain.

28

GENERAL COMMENTS

Medical treatment

The passage in this lesson gives the students an opportunity to reflect on medical treatment in Britain and in their own country. Even if it is unlikely that they should go to Britain, and, one hopes, even more unlikely that they'll need medical treatment, focusing on similarities and differences creates a chance to see how other cultures view illness and their relationship with the doctor.
In some countries, for example, you identify what part of your body is causing you pain or discomfort and you go to a specialist doctor. You very often pay the doctor or the receptionist for any treatment. You may be able to buy the medicine you need at the doctor's surgery. The doctor may not speak to you directly, but in the third person singular, so as to underline the hierarchy in the doctor-patient relationship. In some hospitals you may be obliged to ask family or friends to bring food for you.

Should

Should, like *can* (Lesson 13) and *must* (Lesson 31) is a modal verb. In later levels of *Reward*, more emphasis is given to the particular properties of modal verbs, but at this level, they are treated cursorily. In the Grammar review there is a more complete description of modal verbs, but you want to leave detailed treatment until the Pre-intermediate level.

VOCABULARY AND LISTENING

1 Aim: to present the words and expressions in the vocabulary box.

- If they have them, ask the students to use their dictionaries to do this activity.

- To keep eye contact with the students, you may like to write the four categories on the board, and ask the students to come up, in turn, and write the words under a suitable heading.

Answers
two types of medicine: aspirin, cough medicine
six complaints or illnesses: cold, cough, headache, sore throat, stomach ache, temperature
five adjectives to describe how you feel: dizzy, faint, ill, sick, tired
seven parts of the body: arm, back, finger, foot, hand, leg, toe

2 Aim: to practise listening for specific information.

- Play the tape and ask the students to say what's wrong with each person.

Answers
Conversation 1: headache, cough, feels tired
Conversation 2: stomach ache, headache, temperature
Conversation 3: leg hurts

3 Aim: to present *should*.

- Explain that *should* is used to give advice. Write on the board the answers to activity 2, and ask students to say what each person should do.

FUNCTIONS AND GRAMMAR

1 Aim: to focus on the language of asking and saying how you feel.

- Ask the students to read the information in the functions and grammar box and then to do the activities.

- You may like to write the language in the vocabulary box under *nouns* and *adjectives.*

- Ask the students to work out the answers to the questions by looking at the examples in the vocabulary box.

Answers
1 adjective 2 noun 3 a part of the body

2 Aim: to practise using *should*.

- Ask the students to give their advice about the people in the conversations.

3 Aim: to practise speaking; to practise using the language of asking and saying how people are.

- Ask the students to turn to the relevant Communication activity and act out short conversations.
- When the students are ready, ask them to act out one or two of their conversations in front of the whole class.

READING AND WRITING

1 Aim: to prepare for reading.

- Ask the students to work in pairs and to talk about their answers to the questions.
- You may like to discuss the students' answers with the whole class.

2 Aim: to read for specific information.

- Ask the students to read the passage and to answer the questions in 1.

Answers

1 To the local doctor.
2 Yes.
3 At a chemist's shop.
4 No.
5 Yes, when you're very ill.
6 You call an ambulance on 999.
7 Yes.
8 No, not directly, but through taxes.

3 Aim: to focus on new words.

- Ask the students to read the passage again and to look for the words which mean the same as the statements.

Answers

1 prescription 2 specialist
3 hospital 4 ambulance

4 Aim: to focus on new words.

- Ask the students to check their answers in pairs.
- Correct this activity orally with the whole class.

5 Aim: to practise speaking; to prepare for writing.

- Ask the students to compare medical treatment in Britain and in their countries.

6 Aim: to practise writing.

- Ask the students to prepare an advice leaflet for foreign visitors. Suggest that they use the questions in 1 to help them.
- Ask the students to write a first draft. They should then read out some of their sentences or notes to the whole class. Other students may be able to make suggestions for extra information to be included.
- Ask the students to make a second draft of the written work.
- You may like to ask students to do the last stages of this activity for homework.

The nation's health

In Britain, when you're ill, you go to a doctor near your home. Doctors are men and women, and you can say who you prefer. You usually only spend about ten minutes with the doctor. They can usually say what the matter is very quickly, and often give you a prescription for some medicine. You get this at the chemist's shop. If not, they may suggest you go to a specialist.

Most people only go to their doctor when they're ill. People with colds and coughs don't go to their doctor but to the chemist, to buy medicine. Doctors only come to your home when you're very ill. In an emergency you can call an ambulance on 999. The ambulance takes you to hospital for treatment. Friends and relatives visit you in hospital at certain hours of the day, but they don't stay there.

You don't pay for a visit to the doctor or to the hospital in Britain, but when you work you pay a government tax for your medical care. You also pay for prescriptions if you're over 18.

3 Look at the passage again and find a word which means the same as:

1 A special form for some medicine.
2 A doctor who knows a lot about an illness.
3 A place where there are many people who are ill.
4 A means of transport which takes you to hospital.

4 Work in pairs and discuss your answers to 3.

5 Work in pairs. Is medical care in your country different from Britain?

6 Prepare an advice leaflet about medical care for a foreign visitor to your country. Write answers to the questions in 1. Use the passage to help you.

When you're ill in Japan you should go to a specialist in your illness ...

29 Country factfile

Making comparisons (1): comparative and superlative forms of short adjectives

VOCABULARY AND SOUNDS

1 Match the adjectives with their opposites in the box.

big cold dry fast high hot low old slow small wet young

2 Which adjective does not go with the noun?

1 country hot cold dizzy
2 mountain high big heavy
3 person healthy young high
4 child young low small

3 Which of the words for measurements in the box do you use to describe the following?

area length temperature

centigrade metre millimetre centimetre kilometre square kilometre

4 Listen to the words in the box in 3 and underline the stressed syllable.

Now say the words aloud.

5 Match the words for measurements and their abbreviations.

m mm km cm sq km °C

READING AND LISTENING

1 Look at the country factfile and decide if these statements are true or false.

1 Thailand is smaller than Britain.
2 It's colder in Thailand than in Britain.
3 It's drier in Thailand than in Britain.
4 The population in Thailand is smaller than in Britain.
5 The Thai armed forces are smaller than the British armed forces.
6 Thai children are younger when they start school than British children.
7 Thai children are older when they leave school than British children.

Country factfile

	Kingdom of Thailand	United Kingdom of Great Britain and Northern Ireland	Sweden
1 Land area	513,115 sq km	242,429 sq km	
2 Average temperature	January 25°C July 28°C	January 4.5°C July 18°C	
3 Rainfall	1400 mm	600 mm	
4 Population	(1993) 58,722,437	(1993) 57,970,200	
5 Armed Forces	295,000 troops	274,800 troops	
6 Education	Free and compulsory for all	Free and compulsory for all	

29

GENERAL COMMENTS

Comparative and superlative forms of short adjectives

The adjectives presented in the vocabulary box have been selected on two criteria: they have opposites, and they have endings which illustrate the most common forms of the comparative and superlative. Together, they lend themselves to describing countries more than anything else. So, as in many lessons in the *Reward* series, the topic of this lesson, countries and their statistics, is greatly influenced by the structures which need to be taught.

VOCABULARY AND SOUNDS

1 Aim: to present the adjectives and their opposites.

- Check that the students understand the meaning of the adjectives.
- Ask the students to match the adjectives and their opposites.

Answers
big – small, cold – hot, dry – wet,
fast – slow, high – low, old – young

2 Aim: to focus on collocations.

- Ask the students to choose the odd word out.

Answers
1 dizzy 2 heavy 3 high 4 low

3 Aim: to present the words in the vocabulary box.

- Ask the students to match the measurements and the categories.

Answers
area: square kilometre
length: millimetre, centimetre, metre, kilometre
temperature: centigrade

4 Aim: to focus on stressed syllables in words.

- Play the tape and ask the students to underline the stressed syllables.
- Ask the students to listen and repeat the words.

Answers
centigrade, metre, millimetre, centimetre, kilometre, square kilometre

5 Aim: to focus on abbreviations.

- Ask the students to match the words and the abbreviations.

Answers
m – metre
mm – millimetre
km – kilometre
cm – centimetre
sq km – square kilometre
°C – centigrade

READING AND LISTENING

1 Aim: to practise reading for specific information.

- Ask the students to read the factfile and to decide whether the statements are true (T) or false (F) or whether no information (NI) is given.
- Check this activity with the whole class. Remember that they do not yet have to use comparatives and superlatives.

Answers
1 False 2 False 3 False 4 False
5 False 6 False 7 False

2 Aim: to practise listening for specific information.

- Ask the students to listen to Karl and to put the letter corresponding to the correct answer in the chart.
- Play the tape.

Answers
1 a 2 a 3 a 4 a 5 b 6 b

3 Aim: to practise listening for specific information.

- Ask the students what else Karl mentions.
- Play the tape and ask the students to check their answers.

Answers
Sweden is nearly twice the size of Britain.
January is the coldest month, July the hottest.
It rains less in Stockholm than the national average.
Some students go on to university.

GRAMMAR AND FUNCTIONS

1 Aim: to focus on the formation of comparative and superlative endings.

- Ask the students to read the information in the box and then to do the exercises.
- Ask the students to do this orally with the whole class.

Answers
With *-e* you add *-r, -st.*
With *-y* you drop the *-y* and add *-ier, -iest.*
With vowel + *-d, -g, -m, -n* or *-t* you double the final consonant and add *-er, -est.*

2 Aim: to practise forming comparative and superlative forms of adjectives.

- Ask the students to use the rules in 1 to make comparative and superlative forms.

Answers
large, larger, largest
fine, finer, finest
close, closer, closest
wide, wider, widest
dirty, dirtier, dirtiest
dry, drier, driest
healthy, healthier, healthiest
heavy, heavier, heaviest
noisy, noisier, noisiest
big, bigger, biggest
hot, hotter, hottest
wet, wetter, wettest

3 Aim: to practise using comparatives and superlatives.

- Do this activity orally with the whole class.

Answers
1 Thailand is bigger than Britain.
2 It's hotter in Thailand than in Britain.
3 It's wetter in Thailand than in Britain.
4 The population in Thailand is bigger than in Britain.
5 The Thai armed forces are bigger than the British armed forces.
6 Thai children are older when they start school than British children.
7 Thai children are younger when they leave school than British children.

4 Aim: to practise using comparatives and superlatives.

- Ask the students to do this activity in writing on their own, and then to check their answers in pairs.

Answers
1 Britain is smaller than Sweden.
2 Thailand is hotter than Sweden.
3 Britain is drier than Thailand.
4 Thailand has the biggest armed forces of the three countries.
5 The children in Sweden are older when they start school than the children in Thailand.
6 Sweden is the coldest of the three countries.

WRITING

1 Aim: to practise writing.

- Ask the students to choose one of the tasks in this activity. If they choose to write a factfile, you may like to ask them to do this for homework, using reference books.
- If they choose to write a paragraph, ask them to do a first draft, which they check with you, and then a second, more developed draft. They may like to do this for homework.

2 You're going to hear Karl answering questions about Sweden. Listen and put the letter corresponding to the correct answer in the chart.

1 a 449,964 sq km b 500,200 sq km
2 a January -3°C, July 18°C b January 20°C, July 20°C
3 a 535 mm b 450 mm
4 a 8,730,289 b 31,645,896
5 a 200,500 troops b 64,800 troops
6 a All children between 7 and 16
b All children between 7 and 17

3 Work in pairs. Can you remember what else Karl mentions?

Listen again and check.

GRAMMAR AND FUNCTIONS

Making comparisons (1): comparative and superlative forms of short adjectives

You form the comparative of most short adjectives with *-er*, and the superlative with *-est*.

adjective:	***old***	***large***	***big***	***dirty***
comparative:	***older***	***larger***	***bigger***	***dirtier***
superlative:	***oldest***	***largest***	***biggest***	***dirtiest***

There are some irregular comparative and superlative forms.

good better best
bad worse worst

You use a comparative adjective + *than* when you compare two things which are different

Thailand is ***bigger than*** *Britain.*

1 How do you form comparative and superlative adjectives ending in *-e*, *-y*, vowel + *-d*, *-g*, -m, -n or *-t*?

2 Write the comparative and superlative forms of these adjectives.

large fine close wide dirty dry healthy heavy noisy big hot wet

3 Correct any statements in *Reading and listening* activity 1 which are false.

Thailand is bigger than Britain.

4 Complete these sentences with the adjective in brackets.

1 Britain is _____ than Sweden. (small)
2 Thailand is _____ than Sweden. (hot)
3 Britain is _____ than Thailand. (dry)
4 Thailand has the _____ armed forces of the three countries. (big)
5 The children in Sweden are _____ when they start school than the children in Thailand. (old)
6 Sweden is the _____of the three countries. (cold)

WRITING

Write a factfile for your country. Use the chart in *Reading and listening* activity 1
OR
Write a paragraph comparing your country with Thailand, Britain or Sweden. It doesn't matter if you don't know exact figures.

I think my country is larger than Thailand, but the population is smaller.

30 Olympic spirit

Making comparisons (2): comparative and superlative forms of longer adjectives

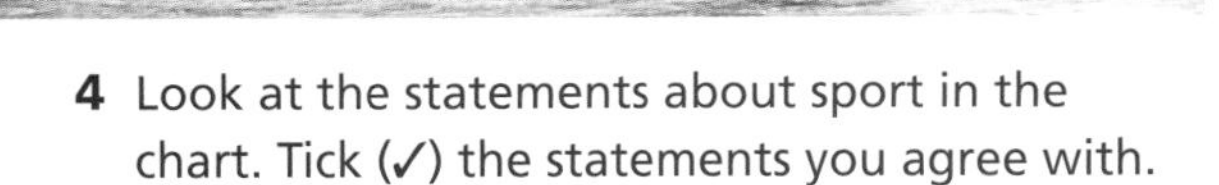

VOCABULARY AND LISTENING

1 Work in pairs. Which of these sports can you see in the photos?

football motor racing swimming
tennis golf horseriding climbing
windsurfing basketball skiing
hang gliding cycling

Turn to Communication activity 28 on page 105 and check you know what the other sports are.

2 Put the words for sports in two columns: *team sports* and *individual sports.*

team sports: football
individual sports: swimming

Now work in pairs and check your answers.

3 Match the adjectives in the box with the sports in the vocabulary box in 1.

popular expensive tiring dangerous
fashionable difficult exciting

football: popular

Now work in pairs and find out if your partner agrees with you.

4 Look at the statements about sport in the chart. Tick (✓) the statements you agree with.

5 Listen to Katy and Andrew talking about their opinions about sport. Tick (✓) the statements they agree with.

6 Work in pairs and check your answers to 5.

Listen again and check.

GRAMMAR AND FUNCTIONS

Making comparisons (2): comparative and superlative forms of longer adjectives

You form the comparative of many long adjectives with *more* + adjective, and the superlative with *most* + adjective.

adjective:	***expensive***	***tiring***
comparative:	***more expensive***	***more tiring***
superlative:	***most expensive***	***most tiring***

*Motor racing is the **most exciting** sport in the world.*
*Climbing is **more difficult** than skiing.*

	You	Katy	Andrew
The most popular sport is football.			
Horseriding is more expensive than cycling.			
Tennis is the most tiring sport.			
Hang gliding is more dangerous than windsurfing.			
Climbing is more difficult than skiing.			

30

GENERAL COMMENTS

Sport

Many students, particularly the male ones, like sport, and teachers may appreciate an opportunity in the textbook for the students to talk about something they really enjoy.

However it's difficult to treat the topic of sport well in a textbook, because it is relatively ephemeral and by the time the material is used in an English class, it will be out of date. This lesson simply presents the words for various sports and creates a neutral context for a popular topic to be discussed.

VOCABULARY AND LISTENING

1 Aim: to present the words in the vocabulary box.

- There are many words for sports which are the same in English as in other languages. Which words do the students recognise?

- Ask the students to decide which sports are shown in the photos.

Answers
windsurfing, cycling, basketball, climbing

2 Aim: to check the comprehension of the new words.

- Ask the students to do this activity on their own and then in pairs.

Answers
team sports: football, basketball, motor racing
individual sports: swimming, tennis, golf, horseriding, climbing, windsurfing, skiing, hang gliding, cycling

- You may find students disagree as to whether motor racing is a team or an individual sport.

3 Aim: to present the new words in the vocabulary box.

- This activity involves the students' personal opinions, so make sure there is plenty of discussion at the feedback stage.

4 Aim: to practise using the new vocabulary; to prepare for listening.

- This activity can be done as discussion in small groups or with the whole class.

- You may like to ask students to raise their hands if they agree with each statement, and then count the number of 'votes'.

5 Aim: to practise listening for main ideas.

- Play the tape and ask the students to tick the statements the speakers agree with.

Answers

	Katy	Andrew
The most popular sport is football.	✓	✓
Horseriding is more expensive than cycling.	✓	✓
Tennis is the most tiring sport.	✗	✓
Hang gliding is more dangerous than windsurfing.	✓	✗
Climbing is more difficult than skiing.	✗	✓
Motor racing is the most exciting sport.	✓	✗

6 Aim: to practise speaking; to provide a second opportunity for listening.

- Ask the students to discuss their answers and then play the tape again.

GRAMMAR AND FUNCTIONS

1 Aim: to practise forming the comparative and superlative forms of adjectives.

- Ask the students to read the information in the grammar and functions box, and then to do the exercises.
- Ask the students to do this activity orally in pairs.
- Correct the answers with the whole class.

Answers

1 Horseriding is very expensive.
Yes, it's the **most expensive** sport I can think of.
2 Motor racing is very dangerous.
Yes, it's **more dangerous** than skiiing.
3 Football is very popular.
Yes, it's **more popular** than tennis.
4 Windsurfing is very difficult.
Yes, it's one of **the most difficult** sports I can think of.
5 Swimming is very tiring.
Yes, it's **the most tiring** sport in the world.

2 Aim: to focus on the formation of comparative and superlative forms of adjectives.

- Ask the students to choose the correct sentence.

Answers

1 a - correct form of superlative.
2 a - only a superlative makes sense.
3 a - *worst* is the superlative form.

3 Aim: to practise speaking.

- Ask the students to work in pairs and to report back on *Vocabulary and listening* activity 4.
- You may like to ask a few students to tell the whole class what their answers were.

4 Aim: to practise forming comparatives and superlatives.

- Ask the students to make sentences using comparative and superlative forms of the adjectives which come from elsewhere in the book.

SPEAKING AND READING

1 Aim: to extend the vocabulary field; to practise speaking.

- Do this activity with the whole class. Write a list of olympic sports on the board.

2 Aim: to practise reading and understanding text organisation.

- Ask the students to read the passage and to put the letters of the paragraphs in the order in which they should go.

Answers

B D A C

3 Aim: to practise reading for main ideas.

- Ask the students to read the passage again and to choose the best title for the story. This activity will encourage them to think of the passage as a whole.

Answer

2

4 Aim: to practise speaking.

- Ask the students to work in groups of two or three and to make a list of what they like or dislike about the Olympic Games.
- Lead a discussion with the whole class about the Olympic Games. How many people enjoy them? How many people dislike them?

1 Complete the sentences using the comparative or superlative form of the adjective.

1 Horseriding is very expensive.
Yes, it's _____ sport I can think of.
2 Motor racing is very dangerous.
Yes, it's _____ than skiing.
3 Football is very popular.
Yes, it's _____ than tennis.
4 Windsurfing is very difficult.
Yes, it's one of _____ sports I can think of.
5 Swimming is very tiring.
Yes, it's the _____ sport in the world.

2 Choose the correct sentence. Can you explain why?

1 a Football is one of the most popular games in the world.
b Football is one of the popularest games in the world.
2 a Skiing is the most difficult sport to do well.
b Skiing is the more difficult sport to do well.
3 a Britain is worse at tennis than many countries.
b Britain is worst at tennis than many countries.

3 Work in pairs. Find out how your partner completed the chart in *Vocabulary and listening* activity 4.

I think hang gliding is the most dangerous sport.
Do you? I think motor racing is more dangerous than hang gliding.

4 Write sentences using the comparative or superlative form of these adjectives.

interesting lively boring intelligent successful enjoyable

The most interesting game in the world is chess.

SPEAKING AND READING

1 With the rest of the class, make a list of Olympic sports.

swimming, athletics, ...

2 Put these paragraphs from a story about the Olympic Games in the right order.

A He then caught the train back to Stockholm, made a reservation into a hotel, got a boat to Japan, got married, had six children and ten grandchildren.

B In 1966 Shizo Kanakuri finished the Olympic marathon in record time. To run the 42 kilometres, he took 54 years, 8 months, 6 days, 8 hours and 32 minutes.

C Then he went back to the place where he stopped for a drink in 1912 and finished the marathon for Japan.

D He started in 1912 in Stockholm, and after a few miles he saw some people having a drink. He was thirsty too, so he joined them.

Adapted from *The Return of Heroic Failures*, by Stephen Pile

3 Choose the best title for the story.

1 The most expensive game 2 The slowest run
3 The worst match

4 Work in groups of two or three. Find out if people in your class enjoy the Olympic Games. If they enjoy them, what do they like and why? If they don't enjoy them, what do they dislike and why?

Do you like the Olympic Games?
What is the most enjoyable game?
What is the most boring game?

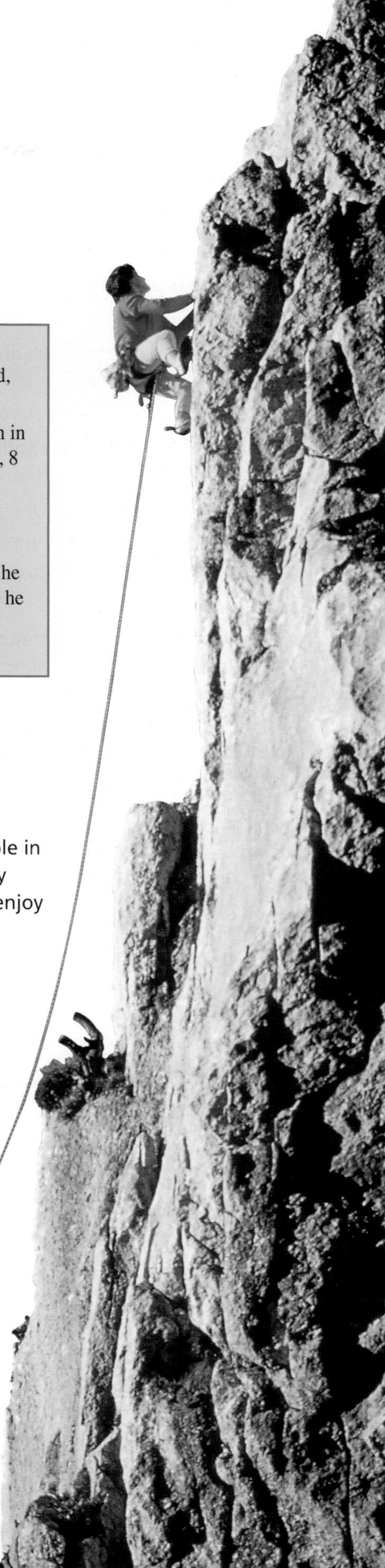

Progress check 26–30

VOCABULARY

1 When you write down an adjective make a note of its opposite meaning.

hot – cold *light – dark*

Match the adjectives and their opposites. (There may be more than one possibility.)

heavy large light long round short small square

You may like to look through Lessons 1 to 30 and see if there are other adjectives and their opposites.

2 Work in groups of three or four and play *Word Zigzag* with words from Lessons 26 to 30.

How to play Word Zigzag

1. On a large sheet of paper, Student A writes a word from Lessons 26 to 30 horizontally.
2. Student B thinks of a word which includes one letter from Student A's word and writes it vertically.
3. Student C thinks of another word which includes a letter from Student B's word and writes it horizontally.
4. The game continues until no one can think of a suitable word. The last student to write a word is the winner.

```
              s t r a i g h t
m o u s t a c h e
              o
              r
              t
```

GRAMMAR

1 Choose the correct word in these sentences.

1 Is this *my/mine* pen? No, it's *my/mine*.
2 Whose are these? They're *her/hers*.
3 Where did you leave *your/yours* coat?
4 This isn't *their/theirs*. It's *your/yours*.
5 Have you got *my/mine* ticket?
6 These are *my/mine* gloves, not *your/yours*.

2 Write the comparative and superlative forms of the following adjectives.

large good popular big ridiculous healthy safe wet expensive tiring heavy high difficult

large larger largest

3 Complete these sentences with the comparative form of the adjective in brackets.

1 France is _____ than the USA. (small)
2 Parachuting is _____ than skiing. (dangerous)
3 Hang gliding is _____ than windsurfing. (expensive)
4 Winter in Norway is _____ than winter in Brazil. (cold)
5 Football is _____ than boxing. (popular)
6 Cycling is _____ than hang gliding. (exciting)

4 Complete the sentences using the comparative or superlative form of the adjective.

1 Russia is a very large country. Yes, it's the _____ country in the world.
2 Motor racing is very expensive. Yes, it's the _____ sport I can think of.
3 Switzerland is quite a small country. Yes, it's _____ than Britain.
4 Swimming is a tiring sport. Yes, it's _____ than cricket.
5 Football is a very popular sport. Yes, it's the _____ sport in the world.

Progress check 26 - 30

GENERAL COMMENTS

You can work through this Progress check in the order shown, or concentrate on areas which have caused difficulty in Lessons 26 - 30. You can also let the students choose the activities they would like or feel the need to do.

VOCABULARY

1 Aim: to present adjectives and their opposites.

- Explain that many adjectives have an opposite, and that it's useful to note it down at the same time as they note down the new word.
- Ask the students to write the opposites of the adjectives shown.

Answers
heavy - light, large - small, long - short, round- square

2 Aim: to revise new words from Lessons 26 to 30.

- Ask the students to play this game in groups of three or four. You can allow up to ten minutes for it.

GRAMMAR

1 Aim: to revise possessive pronouns and adjectives.

Answers
1 my, mine 2 hers 3 your
4 theirs, yours 5 my 6 my, yours

2 Aim: to revise the comparative and superlative forms of adjectives.

Answers

Adjective	Comparative	Superlative
large	larger	largest
good	better	best
popular	more popular	most popular
big	bigger	biggest
ridiculous	more ridiculous	most ridiculous
healthy	healthier	healthiest
safe	safer	safest
wet	wetter	wettest
expensive	more expensive	most expensive
tiring	more tiring	most tiring
heavy	heavier	heaviest
high	higher	highest
difficult	more difficult	most difficult

3 Aim: to revise comparative forms of adjectives.

Answers
1 smaller
2 more dangerous
3 more expensive
4 colder
5 more popular
6 more exciting

4 Aim: to revise the distinction between the comparative and superlative forms of adjectives.

Answers
1 largest
2 most expensive
3 smaller
4 more tiring
5 most popular

5 Aim: to revise the language of saying how you feel.

Answers
1 dizzy, sick, tired
2 leg, arm, back
3 a temperature, a headache, a stomach ache.

6 Aim: to revise the language for giving advice.

- Ask the students to think of suitable advice to give.

SOUNDS

1 Aim: to focus on words which rhyme.

- The spelling of these words does not give much of a clue to their pronunciation. Say the words aloud and ask the students to group them according to those which rhyme.
- Play the tape and ask the students to check, and to repeat each word as they hear it.

Answers
said, head, red
scarf, half, laugh
could, wood

2 Aim: to focus on /eɪ/ and /aɪ/.

- Ask the students to say the words aloud and to put them in two columns.

Answers
/eɪ/: face, May, Spain, maid, tray
/aɪ/: fine, sign, night, lie

- Play the tape and ask the students to check their answers. Ask them to say the words aloud at the same time.

3 Aim: to focus on stress patterns in words.

- Ask the students to match the words and the stress patterns.

Answers
□ ▫ ▫ caravan, camera, temperature
▫ □ ▫ reception, receiver, umbrella
▫ □ ▫ ▫ rectangular

- Play the tape and ask the students to listen and check their answers. As they listen, they should repeat the words.

4 Aim: to focus on word stress in sentences.

- Remind the students that the speakers are likely to stress the words which they consider to be important.

Answers

Customer	And I must have lost it then.
Official	Just say your name again, madam.
Customer	Mary Walter.
Official	And your address and phone number?
Customer	21, Tree Road, Leeds, 75889.
Official	And it was a black plastic bag, you say?
Customer	Yes.
Official	And you last saw it on 20th March at two in the afternoon?
Customer	Yes, in the supermarket.
Official	And what was in it?
Customer	All my shopping and my purse.

5 Aim: to focus on interested intonation.

- Play the tape and ask students to check their answers to activity 4.
- Ask the students to listen to the intonation of the two speakers and to decide who sounds interested.

Answer
The woman sounds more interested than the official.

SPEAKING

1 Aim: to practise speaking.

- Ask the students to work in groups of three or four and to decide where they're likely to hear the sentences.

Possible answers
1 In a chemist's.
2 In a clothes shop.
3 At a Lost Property office.
4 In a doctor's surgery.

2 Aim: to practise speaking.

- Ask the students to match the statements and replies.

Answers
1c 2b 3a 4e

3 Aim: to practise speaking.

- Ask the students to continue one or two of the conversations in writing.
- You may like to ask the students to do this activity for homework.

5 Write two words to complete the following sentences.

1 I feel _____.
2 My _____ hurt/hurts.
3 I've got _____.

6 Reply to these people and give advice. Use *should/shouldn't*.

1 I'm tired.
2 I've got toothache.
3 My back hurts.
4 I feel sick.
5 I've got a cold.
6 I've got a cough.

SOUNDS

1 Group the words which rhyme.

could scarf half said head wood red laugh

Listen and check. As you listen, say the words aloud.

2 Say these words aloud. Is the underlined sound /eɪ/ or /aɪ/?

face fine sign May Spain night lie maid tray

Listen and check. As you listen, say the words aloud.

3 Match the words and the stress patterns.

□▫▫ ▫□▫ ▫□▫▫

reception caravan camera rectangular receiver umbrella temperature

Listen and check. As you listen, say the words aloud.

4 Underline the words you think the speakers will stress.

CUSTOMER	And I must have lost it then.
OFFICIAL	Just say your name again, madam.
CUSTOMER	Mary Walter.
OFFICIAL	And your address and phone number?
CUSTOMER	21, Tree Road, Leeds, 75889.
OFFICIAL	And it was a black plastic bag, you say?
CUSTOMER	Yes.
OFFICIAL	And you last saw it on 20th March at two in the afternoon.
CUSTOMER	Yes, in the supermarket.
OFFICIAL	And what was in it?
CUSTOMER	All my shopping and my purse.

5 Listen and check. Which speaker sounds interested?

Now work in pairs and act out the conversation. Try to sound interested.

SPEAKING

1 Work in groups of three or four. Look at these sentences and decide what the situation is.

1 I've got a sore throat.
2 Have you got it in red?
3 When did you lose it?
4 My back hurts.

2 Match the sentences in 1 with the replies below.

a This morning.
b Not in your size, we haven't.
c Yes, it does look a bit red.
d What have you done?

3 Choose one or two conversations and write a few sentences to continue them. When you're ready, act them out to the rest of the class.

31 When in Rome, do as the Romans do

Needn't, can, must, mustn't

VOCABULARY AND READING

1 Work in pairs. Use the words in the box to say what's happening in the photos.

shake hands   cover   point at   kiss   take off

2 Read *When in Rome* and match the rules and advice with the photos.

When in Rome

- ☐ In parts of Africa you must ask if you want to take a photograph of someone.
- ☐ In Japan you must take off your shoes when you go into someone's house.
- ☐ In Saudi Arabia women must cover their heads in public.
- ☐ In Britain you mustn't point at people.
- ☐ In Japan you mustn't look people in the eye.
- ☐ In China you mustn't kiss in public.
- ☐ In Taiwan you must give a gift with both hands.
- ☐ In France you must shake hands when you meet someone.

31

GENERAL COMMENTS

Needn't, can, must, mustn't

These structures represent a significant learning load for students at this level. However, our research has suggested that a number of teachers expect the structures to be covered at elementary level. Because of the conceptual complexity of the structures (obligation, prohibition, absence of obligation and permission), the main focus of the work is on *must* and *mustn't. Needn't* and *can* are presented almost purely for receptive use.

Rules and advice

The rules and advice are presented in the context of traditions and customs in different cultures. None of them are rules in the strict sense of the term, but pieces of strong advice with the suggestion that not following the advice would be as serious as breaking some kind of unwritten law.

VOCABULARY AND READING

1 Aim: to present the words in the vocabulary box; to pre-teach some difficult words.

- Ask the students to describe the photos using the words in the box.

2 Aim: to practise reading for main ideas.

- The photos constitute a form of summary for the rules and advice. The activity is designed to dissuade the students from trying to understand every single word.

LISTENING

1 Aim: to practise listening for main ideas.

- Explain to the students that they are going to hear an Australian responding to the statements. At this stage in the course the students should be able to cope with more challenging listening passages. In this listening activity, James comments on all the rules and advice mentioned in the passage but the task is graded in such a way that the students are only ticking the things mentioned, in order that they don't feel they must understand every word. You could ask the students if they remember one or two of the differences James mentioned, if you feel they are confident, but this is only a gist listening at this stage.

- Play the tape.

Answers

James talks about all the advice and rules mentioned in the passage but there are differences in Australia.

2 Aim: to present *needn't* and *can*.

- Most of the rules and advice were not applicable to Australia, so *needn't* (absence of obligation) and *can* (permission) can be presented.

- Ask the students to look at the example sentences and then complete the other sentences.

Answers

1 You **needn't** ask if you want to take a photograph of someone.
2 You **needn't** take off your shoes when you go into someone's house.
3 You **can** kiss in public.
4 You **needn't** shake hands with everyone when you meet them in Australia. You **can** shake hands when you meet someone for the first time.

3 Aim: to check comprehension; to provide an opportunity for a second listening.

- Play the tape and ask the students to listen again and to check their answers.

GRAMMAR

1 Aim: to focus on the difference between *must* and *mustn't*.

● Ask the students to read the information in the grammar box and then to do the exercises.

● Ask the students to do this activity orally.

Answers

1 Children **mustn't** play near the road.
2 You **must** be quiet in a library.
3 You **must** keep your wallet in a safe place.
4 Men **must** take off their hats in a church.
5 You **mustn't** give a gift with one had in Taiwan.
6 You **mustn't** wear shoes in a Japanese home.

2 Aim: to practise using *must* and *mustn't* for strong advice.

● Ask the students to work in pairs and to write some advice for language learners.

● Ask the students to place their advice where the other students can see it.

3 Aim: to practise using *must* and *mustn't* for rules.

● You may like to ask the students to do this orally in groups.

● Ask each group to tell the rest of the class the rules they have discussed for the school or their place of work.

SOUNDS

1 Aim: to focus on the pronunciation of *must* and *mustn't* in connected speech.

● Play the tape and point out that *must* and *mustn't* are pronounced /mʌst/ and /mʌsnt/.

2 Aim: to practise pronouncing *must* and *mustn't* in connected speech.

● Ask the students to repeat the sentences. Make sure they pronounce *must* and *mustn't* correctly.

SPEAKING AND WRITING

1 Aim: to use the target structures; to practise speaking.

● Ask the students to work in pairs and to think of strong advice and rules they can give to visitors to their country.

2 Aim: to practise writing and speaking.

● To recycle the language covered in 1, ask the students to work with another pair and to draw up a list of advice and rules.

● You may like to ask the students to do the written part of this activity for homework.

LISTENING

1 Listen to James, who's Australian, talking about some of the advice and rules in *When in Rome*. Tick (✓) the statements he talks about.

2 Look at these sentences.

Women needn't cover their heads in Australia.

In Australia you can look people in the eye.

Complete these sentences with *needn't* and *can* so that they are true for Australia.

1. You _____ ask if you want to take a photograph of someone.
2. You _____ take your shoes off when you go into someone's house.
3. You _____ kiss in public in Australia.
4. You _____ shake hands with everyone when you meet them in Australia. You _____ shake hands when you meet someone for the first time.

3 Now listen again and check.

GRAMMAR

Needn't and can

You use *needn't* if it isn't necessary to do something.

*Women **needn't** cover their heads in Australia.*

You use *can* if you're allowed to do something.

*In Australia you **can** look people in the eye.*

***Can* is also a modal verb. (For more information about modal verbs see Grammar review page 111.)**

Must, mustn't

***Must* and *mustn't* are modal verbs.**

You use *must* to talk about something you're strongly advised to do or are obliged to do, such as rules.

*You **must** ask if you want to take a photograph of someone.* (strong advice)

*Women **must** cover their heads in public.* (rule)

You use *mustn't* to talk about something you're strongly advised not to do or are not allowed to do.

*You **mustn't** point at people.* (strong advice)

*You **mustn't** kiss in public.* (rule)

1 Complete these sentences with *must* or *mustn't*.

1. Children _____ play near the road.
2. You _____ be quiet in a library.
3. You _____ keep your wallet in a safe place.
4. Men _____ take off their hats in a church.
5. You _____ give a gift with one hand in Taiwan.
6. You _____ wear shoes in a Japanese home.

2 Write some strong advice for people learning a foreign language.

You must come to classes every week.
You mustn't miss any lessons.

3 Write some rules for your school or the place where you work.

You mustn't smoke during lessons.

SOUNDS

1 Listen to the way you pronounce *must* /mʌst/ and *mustn't* /mʌsnt/. Look at the sentences in *Grammar* activity 1.

2 Say the sentences aloud. Make sure you pronounce *must* and *mustn't* correctly.

SPEAKING AND WRITING

1 Work in pairs. Think of strong advice and rules you can give to visitors to your country about the following:

– giving presents
– what to wear
– home visits
– eating habits
– table manners

When you receive a gift in Spain, you must open it immediately.

2 Work with another pair. Do you have similar rules and advice? Write a list of advice and rules for visitors to your country.

32 *Have you ever been to London?*

Present perfect (1): talking about experiences

READING AND VOCABULARY

1 Work in pairs. Make a list of famous sights to see in your town or capital city.

2 Look at these famous sights of London. Do you know what they are?

3 Read this postcard from London. Which photo in 2 is on the back of it?

4 Here is a list of some of the things you can do in London. Tick (✓) the places Guy and Emma have been to or the things they've done.

- ☐ watch the Changing of the Guard
- ☐ visit Westminster Abbey
- ☐ visit St Paul's Cathedral
- ☐ listen to a concert in St James' Park
- ☐ climb Tower Bridge
- ☐ go to Hampstead

Dear David and Anna,

Hi! How are you? We're having a wonderful time in London. We're staying in a hotel in the centre of London. We've only been here four days but we've done so much already. We've watched the Changing of the Guard at Buckingham Palace and we've listened to a concert in St James' Park. We've visited St Paul's Cathedral, but not Westminster Abbey. We've climbed Tower Bridge (you can see it on this postcard) and we've been to Greenwich by boat, but we haven't been to Hampstead yet. We're going there tomorrow.

See you soon!

Love Guy and Emma

David and Anna Sayle
Apartment 214
51st West City Street
New York 10021
USA

32

GENERAL COMMENTS

Present perfect tense

This is the first of two lessons in *Reward* Elementary on the present perfect tense. Although it has an equivalent in form in a number of European languages, it is used differently. This lesson focuses on its use to talk about experience.

It should be noted that in American English, and in some contexts in British English, such as newspaper reports, the past simple is used where one might expect the present perfect. *Did you speak to your mother yet?*

It may be that we are witnessing a shift in usage away from the present perfect.

READING AND VOCABULARY

1 Aim: to practise speaking.

- The lesson is about visiting London and other cities with sights to see. Ask the students to make a list of the things to see in their own nearest city.

2 Aim: to prepare for reading.

- Ask the students if they have been to London. If they have, do they recognise the places in the photos? Can they think of anything they'd like to see if they went to London?

3 Aim: to practise reading for main ideas; to present the present perfect.

- Ask the students to read the postcard and to say which photo is on the back of it.

Answer

Tower Bridge

4 Aim: to check comprehension; to practise reading for specific information.

- Ask the students to read the postcard again and to tick the places Guy and Emma have been to or the things they've done.

Answers

✓ watch the Changing of the Guard
visit Westminster Abbey
✓ visit St Paul's Cathedral
✓ listen to a concert in St James' Park
✓ climb Tower Bridge
go to Hampstead

5 Aim: to present the words in the vocabulary box.

- Check that the students understand what the words mean. They have already occurred in the lesson, and some of them elsewhere in *Reward* Elementary.

Answers

1 A cathedral is the most important church in a city.
2 You cross a river by going over a bridge or by taking a boat.
3 When you climb something, you go up it.
4 A park is a public place in a town, with trees and grass.
5 The view from a building is what you can see from it.

SOUNDS

1 Aim: to focus on the pronunciation of the present perfect.

- Play the tape and ask the students to repeat the phrases.

2 Aim: to focus on stressed words in sentences.

- Play the tape and ask the students to read and listen to the conversation. As they listen, they should underline the stressed syllables. You may need to play the tape several times.

- You can continue this activity by asking the students to write on a separate piece of paper all the words they underlined.

- Then when they're ready, and without looking at their books, they should reconstitute the conversation using the key words as prompts.

Answers

A Have you ever been to London?
B No, I haven't. I've never been there.
A Have you ever stayed in a hotel?
B Yes, I have.
A When was that?
B When I was in Spain last year.

3 Aim: to practise speaking.

- Ask the students to practise the conversation in pairs.

- You may like to ask the students to perform the conversation in front of the class.

GRAMMAR

1 Aim: to practise talking about experiences using the present perfect.

- Ask the students to read the information in the grammar box, and then to do the exercises.

- Ask the students to work in pairs and to ask and answer questions about Guy and Emma's holiday. Refer the students back to their answers in *Reading and vocabulary*, activity 4.

2 Aim: to focus on the form of regular past participles.

- Ask the students to look at the past participles and to write the infinitives.

Answers

live, work, stay, watch, visit, listen

3 Aim: to focus on some irregular past participles.

- Ask the students to do this activity orally in pairs.

Answers

eat - eaten, drink - drunk, drive - driven, read - read, see - seen, fly - flown, take - taken, buy - bought, win - won, make - made, write - written, send - sent

SPEAKING AND WRITING

1 Aim: to practise speaking.

- Ask the students to think about a list of places in the town where they are now. Ask them to go round the class asking and saying what they've seen or done in the town.

2 Aim: to practise writing.

- Ask the students to work alone and to write a postcard to a friend. Ask them to do a first draft.

- Ask the students to show you their first drafts. Make suggestions for improvements and then ask the students to do a second draft.

- You may like to ask students to do this activity for homework.

5 Complete the sentences below with the words in the box.

climb cathedral bridge boat park view

1 A _____ is the most important church in a city.
2 You cross a river by going over a _____ or by taking a _____.
3 When you _____ something, you go up it.
4 A _____ is a public place in a town, with trees and grass.
5 The _____ from a building is what you can see from it.

SOUNDS

1 Listen and repeat.

ever Have you ever Have you ever been
Have you ever been to London?
ever Have you ever Have you ever stayed
Have you ever stayed in a hotel?

2 Read and listen to this conversation. Underline the stressed words.

A Have you ever been to London?
B No, I haven't. I've never been there.
A Have you ever stayed in a hotel?
B Yes, I have.
A When was that?
B When I was in Spain last year.

3 Work in pairs and practise the conversation in 2.

GRAMMAR

Present perfect (1): talking about experiences

You use the present perfect to talk about an experience, often with *ever* and *never*.

Have you ever stayed *in a hotel?*

(= Do you have the experience of staying in a hotel?)

Yes, I have. (= Yes, at some time in my life, but it's not important when.)

No, I haven't. ***I've never stayed*** *in a hotel.*

You form the present perfect with *has/have* + past participle. You usually use the contracted form, *'ve* or *'s*.

Many past participles are irregular.

Have you ever been to London? We've done so much.

For a list of irregular participles, see page114.

Remember that you use the past simple to talk about a definite time in the past.

When did you stay in a hotel? When I was in Spain last year.

1 Work in pairs. Ask and answer questions about what Guy and Emma have done on their holiday.

1 watch the Changing of the Guard
2 visit St Paul's Cathedral
3 climb Tower Bridge
4 visit Westminster Abbey
5 listen to a concert in St James' Park
6 go to Hampstead

1 Have they watched the Changing of the Guard?
Yes, they have.

2 Here are the regular past participles of some verbs. Write the infinitive.

lived worked stayed watched visited listened

3 Match the infinitives with their irregular past participles.

eat drink drive read see fly take buy win make write send

driven sent read seen flown bought won made eaten drunk taken written

SPEAKING AND WRITING

1 Think of things to do and places to see in the town where you are now. Go round the class and ask and say what people have done in your town.

Have you seen the cathedral? Yes, I have.
Have you taken the boat along the river?
No, I haven't.

2 Imagine you're a visitor to your town. Write a postcard to a friend saying what you've seen and where you've been. Use the postcard in *Reading* activity 3 to help you.

Dear Enrique,

Hi, how are you? I'm in Seville at the moment. I've seen the Alcazar...

33 What's happened?

Present perfect (2): talking about recent events; *just* and *yet*

LISTENING AND VOCABULARY

1 Match these sentences with the pictures of Barry.

He's hurt his back.   He's lost his wallet.   He's failed his exam.

2 Look at this conversation. Can you guess what Barry says?

ALAN Hi! How's your day been?
BARRY (1) _____
ALAN I'm sorry to hear that. What's happened?
BARRY (2) _____
ALAN Your back! How did you hurt it?
BARRY (3) _____
ALAN A box of books! I'm not surprised you hurt yourself trying to lift a box of books. Have you been to the doctor yet?
BARRY (4) _____
ALAN Well, I think you should go immediately. And what else has happened?
BARRY (5) _____
ALAN Your wallet? Where did you lose it?
BARRY (6) _____
ALAN Have you been back to the bus stop yet?
BARRY (7) _____
ALAN And have you heard your exam result?
BARRY (8) _____
ALAN Have you passed?
BARRY (9) _____
ALAN Oh dear, it's been one of those days for you, hasn't it?

3 Listen and check your answers to 2.

4 Work in pairs and act out the conversation.

5 Match the verbs and the nouns in the box. There may be several possibilities.

break cut plate goal drop bag miss wallet lose pass fail
arm pay score steal bill exam finger train crash car catch

break – arm, finger...

33

GENERAL COMMENTS

Present perfect

You may find that the most useful source of present perfect tenses in everyday speech is to be found in radio news reports and newspapers. You may not think your students are ready to listen to the radio or read newspapers yet, but with carefully designed activities, you'll find they'll be able to understand a little of what they hear or read.

You may want to consider the possibility of introducing the radio or newspapers into the classroom. Simple activities in which they listen and read and note down any words which they recognise are useful: they are likely to notice proper names, countries, geographical regions, international words, occupations and more. They will also be receiving some exposure to real-life English in the reassuring context of the classroom.

You can find out times of BBC World Service radio programmes and wavelengths by writing to BBC, Bush House, London, W1 England.

LISTENING AND VOCABULARY

1 Aim: to present the present perfect for talking about recent events.

- These sentences show the target structure and its meaning. Ask the students to match the sentences to the pictures of Barry.

2 Aim: to prepare for listening; to present the present perfect for talking about recent events.

- The conversation shows more examples of the recent events which have happened to Barry. Ask the students to read the passage and to guess what he says.
- Ask the students to check their answers in pairs.

3 Aim: to practise listening.

- Play the tape and ask the students to listen and check their answers to 2.
- Find out how many students got correct answers.

Answers

1 Awful, absolutely awful.
2 Well, I've hurt my back.
3 I tried to lift a box of books.
4 No, not yet.
5 I've lost my wallet.
6 At the bus stop, I think.
7 No, I haven't.
8 Yes, I have.
9 No, I've failed it.

4 Aim: to practise saying the target language.

- Ask the students to work in pairs and act out the conversation.

5 Aim: to present the words in the vocabulary box.

- These words are used not only in the conversation you've just played but in the *Speaking* activities at the end of the lesson. Ask the students to match the verbs with the nouns.

Answers

break: plate, arm, finger
cut: arm, finger
drop: plate, wallet, bag
miss: goal, train
lose: wallet, bag
pass: exam
fail: exam
pay: bill
score: goal
steal: bag, wallet
crash: car
catch: train

6 Aim: to practise listening for specific information.

- Ask the students to listen to four more conversations about events which have just happened.
- Play the tape and ask the students to put the correct conversation number alongside the verbs in 5 which they hear.

Answers

Conversation 1: steal
Conversation 2: miss
Conversation: 3 score
Conversation 4: cut

GRAMMAR

1 Aim: to practise using the present perfect for recent events.

- Ask the students to read the information in the grammar box and then to do the exercises.

- Ask the students to think about the conversations they heard and to say what has happened.

Answers
1 Someone has stolen her bag.
2 They've just missed the train.
3 He's scored a goal.
4 She's cut her finger.

2 Aim: to focus on the position of *just* and *yet* in present perfect sentences.

- Ask the students to look at the example sentences in the grammar box and to say where *just* and *yet* go in the sentence.

Answer
just: between the auxiliary and the past participle.
yet: at the end of the sentence.

3 Aim: to check the use of the present perfect tense.

- Ask the students to do this activity on their own and then to check their answers in pairs.

Answers
1b 2a 3b 4a 5a 6b

- Spend some time explaining why these answers are correct.

SPEAKING

1 Aim: to prepare for activity 2.

- Ask the students to work on their own for this activity.

2 Aim: to practise using *yet* and the present perfect.

- Ask the students to work in pairs and to ask and say what they've done or haven't done yet.

- You may like to ask some students to do this activity in front of the whole class.

3 Aim: to practise using the target structures.

- Ask the students to act out the situations in pairs.

- Ask the students to choose three or four of the situations and write the conversations. Ask them to extend the conversation as much as possible.

- Ask the students to perform one or two of their conversations with the rest of the class.

- You may like to ask the students to do the written part of this activity for homework.

6 Listen to four more conversations. Put the number of the conversation by the verbs in 5 which you hear.

GRAMMAR

Present perfect (2): talking about recent events; *just* and *yet.*

You use the present perfect to talk about recent events, such as a past action which has a result in the present. You often use to it describe a change.

*He's **hurt** his back.* *He's **lost** his wallet.*

You use *just* if the action is very recent.

*He's **just** lost his wallet.*

You use *yet* in questions and negatives to talk about an action which is expected.

*Have you been to the doctor **yet?***

*I haven't gone back to the bus stop **yet.***

Remember that you use the past simple to say when the action happened.

When did you lose your wallet? I lost it this morning.

1 Write sentences saying what has happened in the conversations in *Listening and vocabulary* activity 6.

In the first conversation, someone has stolen her bag.

2 Look at the example sentences in the grammar box. Where do you put *just* and *yet* in a sentence?

3 Choose the correct sentence. Can you explain why?

1 a I *'ve seen* her yesterday.
  b I *saw* her yesterday.
2 a *Have you been* shopping yet?
  b *Did you go* shopping yet?
3 a She*'s lost* her wallet last week.
  b She *lost* her wallet last week.
4 a I*'ve never been* to England.
  b I *never went* to England.
5 a We*'ve just decided* where to go on holiday.
  b We *just decided* where to go on holiday.
6 a *Have you met* anyone famous when you were in Hollywood?
  b *Did you meet* anyone famous when you were in Hollywood?

SPEAKING

1 Make a list of things you've done and things you haven't done yet this week.

call my mother, pay the bills, do some shopping, write to the bank manager, . . .

2 Work in pairs. Show each other the lists you made in 1. Ask and say what you've done and haven't done yet.

Have you called your mother yet?
Yes, I have.
Have you paid the bills yet?
No, I haven't.

3 Work in pairs. Act out the following situations.

Student A: Look at the statements below and tell Student B what's happened using the present perfect. Answer his/her questions.

– you break your arm
– you find some money
– you lose your friend's pen
– you fail your exam
– you drop an expensive plate

I've broken my arm!

When you've finished, react to Student B's situations. Ask questions using the past simple.

Student B: React to Student A's situations. Ask questions using the past simple.

STUDENT A *I've broken my arm!*
STUDENT B *How did you do that?*
Where did it happen?

When you've finished, tell Student A what's happened to you. Remember to use the present perfect tense.

– you win a million pounds
– you cut your finger
– you miss the last bus home
– you burn your dinner
– you crash the car

Planning a perfect day

Imperatives; infinitive of purpose

SPEAKING

1 Work in pairs. What is your idea of a perfect day out? Here are some suggestions:

- a shopping trip
- a visit to the beach
- a picnic in the country
- a visit to a historical building
- a visit to some friends
- a walk in the mountains

2 Look at the photo. Which of the situations in 1 does it show?

4 Complete these sentences with words from the passage.

1 I was very _____ so I had something to eat.
2 The weather was _____ so we had dinner outside.
3 When you go away for a night or two, don't forget to _____ your toothbrush.
4 Check the weather _____ to find out if it's going to rain.
5 I was cold in bed so I asked for another _____.

5 Work in pairs. Do you agree with the advice in the passage? Do you have picnics like this in your country?

READING AND VOCABULARY

1 Work in pairs. Make a list of things to do or take on a perfect picnic.

2 Read *The perfect picnic* and decide which paragraphs the pictures illustrate.

3 Match these words with the pictures.

bottle opener barbecue matches knife fork ice cup carton rubbish blanket

34

GENERAL COMMENTS

Imperatives

Students will have come across imperatives if they have followed *Reward* Starter, and they will have had passive exposure to imperatives in all the instructions in *Reward* Elementary. So the structure shouldn't cause the students a great deal of difficulty and for this reason, there are no specific activities in the *Grammar* section.

Picnics

The word *picnic* comes from the French *pique-nique* and was originally used to describe a fashionable social entertainment in which everyone brought food to be shared. Nowadays, it is understood to be a meal, often of cold food, to be eaten outdoors.

A picnic may be an activity which your students do not do very often, if at all, in their culture. If not try to choose an equivalent activity, perhaps outdoors, which involves some sharing of food with friends and family for discussion.

SPEAKING

1 Aim: to introduce the theme of the lesson; to practise speaking.

- Ask the students to discuss their idea of a perfect day out. The situations in the textbook are only suggestions.

2 Aim: to practise speaking.

- Ask the students to decide which situation from activity 1 they can see in the photo.

Answer
a picnic in the country

READING AND VOCABULARY

1 Aim: to prepare for reading.

- Ask the students to work in pairs and to make a list of things to take on a picnic.
- Ask the students to read out their list. Note down their list of things on the board.

2 Aim: to practise reading for main ideas.

- The drawings represent a summary of each paragraph. If you ask them to do this task, they will be less likely to demand an explanation for difficult vocabulary.

Answers
paragraph 3: rubbish
paragraph 4: knife, fork
paragraph 5: carton of orange juice, cups
paragraph 6: blanket
paragraph 7: ice
paragraph 9: bottle opener, barbeque, matches, knife

3 Aim: to present the words in the vocabulary box.

- The students may not know some of these words. Ask them to try and match the words they know. They will understand the meaning of the other words by a process of elimination.

4 Aim: to present new words from the passage; to practise reading for specific information.

- These words are not considered essential to learn productively, but they may be worth learning for receptive use.

Answers
1 hungry 2 perfect 3 pack
4 forecast 5 blanket

- The students may find this activity quite hard, but the activity type needs to be introduced.

5 Aim: to practise speaking.

- This activity acknowledges the possibility that students may not have picnics in their country.
- Find out if the kind of picnic described in the passage appeals to the students. Ask them to say where they'd like to have a picnic.

GRAMMAR

1 Aim: to focus on the infinitive of purpose.

- Ask the students to read the information in the grammar box and then to do the exercises.

- Ask the students to read the passage again and find the answers to these comprehension check questions.

Answers
1 **To be** sure there's plenty to do when you finish your picnic.
2 **To avoid** taking knives and forks.
3 **To sit** on or **to keep** you warm.
4 **To keep** it cool.
5 **To allow** everyone to carry something.
6 **To make** people hungry.

2 Aim: to practise using the infinitive of purpose; to check comprehension of new vocabulary.

- Ask the students to do this on their own.

Answers

1 fork	2 barbecue	3 knife
4 ice	5 matches	6 blanket

3 Aim: to practise using the infinitive of purpose; to check comprehension of new vocabulary.

- Ask the students to check their answers in pairs.

- Check the answers orally with the whole class.

4 Aim: to practise using the infinitive of purpose.

- Do this activity with the whole class.

Possible answers
1 To buy food or clothes.
2 To relax and visit a foreign country.
3 To catch a plane or to meet people.
4 To do your job/to learn something and pass your exams.
5 To see a film.
6 To cut something.

WRITING

1 Aim: to practise speaking; to prepare for writing.

- Ask the students to work in groups of two or three. They should decide on one of the suggestions in *Speaking* activity 1. When they have made their decision, they should work alone and make notes on pieces of advice for their chosen outing.

2 Aim: to practise speaking; to practise writing.

- Ask the students to put all their ideas together. Go round the groups giving extra ideas.

- Ask the students to write their advice in full sentences.

- You may like to ask the students to do this writing activity for homework.

3 Aim: to practise speaking.

- Ask the students to show their advice to other students and to discuss it.

- If it's suitable, you may like to put their advice on the wall for everyone to see.

The perfect picnic

Everyone says that food and drink taste better when you have a picnic. But what do you do to have a perfect picnic? Here's some advice.

1 Choose where you want to go very carefully. In the country? In the city? The picnic site should be attractive and interesting, to be sure there's plenty to do when you finish your picnic.
2 Check the weather forecast the day before you go. The perfect picnic needs perfect weather.
3 Don't take too much to carry. For the perfect picnic you leave home with food and drink and you return only with rubbish.
4 Choose small items of food, such as eggs or sandwiches, to avoid taking knives and forks. To make it the perfect picnic, take food which you don't usually eat.
5 Take small cartons of juice or plastic bottles of water. They're more expensive, but they aren't as heavy as glass bottles, cups and glasses.
6 Pack a blanket to sit on or, if it's cold, to keep you warm.
7 Put fresh food in a bag with ice to keep it cool.
8 Put the whole picnic in a number of small bags, to allow everyone to carry something.
9 Prepare everything before you go OR make sure you've got everything you need to finish preparing the picnic, such as a knife, a bottle opener, barbecue, matches.
10 Check there is a short walk to the picnic site to make people hungry.

GRAMMAR

Imperatives

You use an imperative (infinitive without *to*) to give instructions and advice.

***Check** the weather forecast.*

You use *don't* + imperative for a negative instruction.

***Don't take** too much to carry.*

Infinitive of purpose

You use *to* + infinitive:

– to say why people do things.

*Try to have your picnic on a weekday, **to avoid** the weekend traffic.*

– to say what you use something to do.

*You use a bottle opener **to open** bottles.*

1 Answer the questions. Use *to* + infinitive.

1 Why should the picnic site be attractive and interesting?
2 Why should you choose small items of food?
3 Why should you pack a blanket?
4 Why should you put fresh food in a bag with ice?
5 Why should you put the whole picnic in a number of small bags?
6 Why should you check there's a short walk to the picnic site?

1 To be sure there's plenty to do when you finish your picnic.

2 Match words in the vocabulary box in *Reading and vocabulary* activity 3 with what you use them to do.

1 to put food in your mouth
2 to cook food outdoors
3 to cut food
4 to keep something cold
5 to light a barbecue
6 to sit on or to keep you warm

3 Work in pairs and check your answers to 2.

You use a fork to put food in your mouth.

4 Think of reasons why you do the following. Use *to* + infinitive.

1 go shopping
2 take a holiday
3 go to the airport
4 go to work/school
5 go to the cinema
6 use a knife

1 You go shopping to buy food or clothes.

WRITING

1 Work in groups of two or three. You're going to prepare some advice for planning the perfect day out. Make sure you all choose the same situation from *Speaking* activity 1.

First, work alone. Make notes on your advice.

a shopping trip – make a list

2 Work with the rest of the group. Make a list of all your advice, and explain why. Use an imperative and *to* + infinitive.

Make a list of things to buy to be sure you don't forget anything.

3 Show your instructions to another group. Do they agree with your advice? Is there anything which surprises them?

35 She sings well

Adverbs

VOCABULARY AND SOUNDS

1 Match the adverbs and their opposites in the box below.

badly carefully carelessly quickly loudly politely quietly rudely slowly well

2 Listen to four conversations. Which adverbs would you choose to describe how the speakers are speaking?

READING AND SPEAKING

1 Work in pairs. Look at the photos. What do you think the people do?

2 Work in pairs. Who do you think was good at school?

3 Work in pairs. Read the extracts from school reports and match the adult in the photo with the child in the report.

Antonia

Music	She sings well and she plays the piano and guitar beautifully.
Art	She draws very carefully. Her work is excellent.
General	She is a very artistic young woman.

Guillaume

Sport	He can run quickly, and plays tennis well. He's good at most sports.
English	He writes very slowly and his spelling is very bad.
General	He finds many subjects very difficult. Must try harder.

Beate

Maths	She can add and subtract numbers quickly in her head.
Science	She is good at biology. She passed the exam easily.
General	Beate shows great ability in her work.

Kate

French	She speaks French almost fluently. Well done!
English	She writes English compositions confidently. She works very hard.
General	Her manners are excellent. She talks to people quietly and politely.

4 Say what the people were good at when they were at school.

Antonia was good at music.

5 Work in pairs and say what you were good at when you were at your first school.

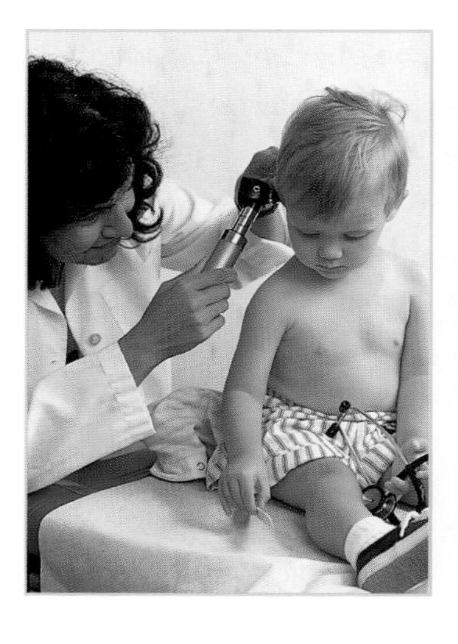

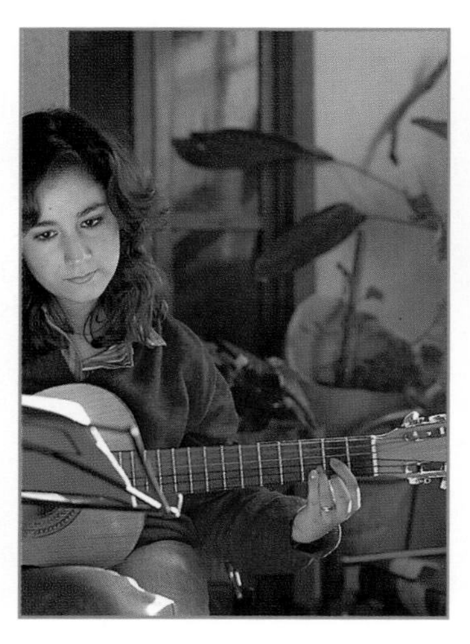

35

GENERAL COMMENTS

Adverbs

The students will already have come across some adverbs of time, such as *often* and *sometimes*, in Lesson 11. This lesson gives a very brief introduction to adverbs of manner. There are many other issues concerning adverbs which are not covered in *Reward* Elementary: the position of adverbs of time, manner and place; adjectives which look like adverbs (such as *friendly*), full coverage of adverbs with the same form as the adjective (*fast, high, low, wide*). These are all covered in later levels of the *Reward* series.

School

This lesson has school as its theme, which lends itself to showing adverbs of manner in a suitable context. If your students are still at school, they should be encouraged to think about their early schooldays and to see if there is a difference in their attitude and behaviour between then and now.

The theme of the lesson concerns school performance and later success in life.

VOCABULARY AND SOUNDS

1 Aim: to present the words in the vocabulary box.

- Ask the students to match the adverbs with their opposites. You may like to do this activity orally with the whole class.

Answers
badly – well, carefully – carelessly, quickly – slowly, loudly – quietly, politely – rudely

2 Aim: to check comprehension of the new words; to practise listening to tone of voice.

- Play the tape and ask the students to say how the four people are speaking.

Answers
Conversation 1: quietly
Conversation 2: rudely
Conversation 3: politely
Conversation 4: loudly

READING AND SPEAKING

1 Aim: to prepare for reading; to practise speaking.

- Ask the students to look at the photos and to say what the people do.

Answers
doctor
musician
sportsman
business woman

2 Aim: to prepare for reading.

- Ask the students to say which of the people in the photos they think were good students at school.

3 Aim: to practise reading for main ideas.

- Tell your students that the reading passage is made up of extracts from school reports. Do the extracts look like their own school reports?

- Ask the students to read the school reports and to match the adult in the photo with the child in the report.

Answers
first photo: Beate
second photo: Antonia
third photo: Guillaume
fourth photo: Kate

4 Aim: to check comprehension.

- Ask the students to say what each student was good at.

Answers
Antonia was good at music and art.
Guillaume was good at sport.
Beate was good at maths and biology.
Kate was good at French and English.

5 Aim: to practise speaking.

- Ask the students to talk about the subjects they were good at. Are they still good at these subjects?

GRAMMAR

1 Aim: to focus on the formation of adverbs.

● Ask the students to read the information in the grammar box and then to do the exercises.

● Do this activity with the whole class. Write the answers on the board.

Answers
badly - bad, carefully - careful, carelessly - careless, quickly - quick, loudly - loud, politely - polite, quietly - quiet, rudely - rude, slowly - slow, well - good

2 Aim: to focus on the formation of adverbs.

● Ask the students to do this activity on their own and then correct it with the whole class.

Answers
angrily, happily, gently, quickly, immediately, successfully, comfortably, suddenly, funnily, frequently

3 Aim: to focus on the distinction between adjectives and adverbs.

● Remind the students that an adverb is used to describe a verb, an adjective is used to describe a noun.

● Ask the students to complete the sentences on their own.

Answers
1 He spoke **clearly** so everyone could hear him **well**.
2 They were late so they had a **quick** game of tennis and then left.
3 She had a very **successful** lesson with her pupils.
4 He listened to his teacher very **carefully**.
5 Could you speak more **slowly** please. Your accent is hard to understand.
6 He passed the spoken exam very **easily**.

4 Aim: to practise speaking; to practise using adverbs.

● Ask the students to talk about what other students can do and how well they can do it.

LISTENING AND SPEAKING

1 Aim: to prepare for listening.

● Ask the students to read the questions about school and to think about their answers.

● You may like to ask a few students for their feedback.

2 Aim: to practise listening for main ideas.

● Play the tape and ask the students to listen and to tick the statements the speakers say *yes* to.

● Explain that the students may not understand every word, but they should understand enough to perform the task.

3 Aim: to practise speaking; to check comprehension.

● Ask the students to check their answers in pairs. Ask them to try to remember exactly what the speakers say.

● Play the tape again.

Answers

	Gavin	Jenny
Do/did you always work very hard?	✓	✓
Do/did you always listen carefully to your teachers?	✓	✗
Do/did you always behave very well ?	✗	✓
Do/did you pass your exams easily?	✗	✗
Do/did you always write slowly and carefully?	✓	✓
Do/did you think schooldays are/were the best days of your life?	✗	✗

4 Aim: to practise speaking.

● Ask the students to ask and answer the questions and to tick the statements they say *yes* to.

● You may like to ask some students what their answers were.

5 Aim: to practise speaking; to practise using adverbs.

● Begin this activity by thinking of an adverb and then saying something in the manner of the adverb. Ask the students to guess the adverb you thought of. Do this two or three times.

● Ask the students to work in groups of three or four and to follow the instructions.

GRAMMAR

> Adverbs
> **You use an adverb to describe a verb.**
> *She speaks* ***slowly.***
> **You usually form an adverb by adding *-ly* to the adjective.**
> ***quiet – quietly***
> ***loud – loudly***
> **If the adjective ends in *-y*, you drop the *-y* and add *-ily*.**
> *easy – easily*
> **Some adverbs have the same form as the adjective they come from.**
> ***late early hard***
> **The adverb from the adjective *good* is *well*.**
> *She's a* ***good*** *singer. She sings* ***well.***

1 Write the adjectives which the adverbs in the vocabulary box come from.

badly – bad

2 Write the adverbs which come from these adjectives.

angry happy gentle quick immediate successful comfortable sudden funny frequent

3 Complete the sentences with the adjective in brackets or the adverb which comes from it.

1 He spoke _____ so everyone could hear him _____. (clear, good)
2 They were late so they had a _____ game of tennis and then left. (quick)
3 She had a very _____ lesson with her pupils. (successful)
4 He listened to his teacher very _____. (careful)
5 Could you speak more _____, please. Your accent is _____ to understand. (slow, hard)
6 He passed the spoken exam very _____. (easy)

4 Work in pairs. Say what people in your class can do and how well they do it. Use an adverb.

Guido can run quickly.

LISTENING AND SPEAKING

1 Think about your answers to these questions about school.

	You	Gavin	Jenny
Do/did you always work very hard?			
Do/did you always listen carefully to your teachers?			
Do/did you always behave very well?			
Do/did you pass your exams easily?			
Do/did you always write your homework slowly and carefully?			
Do you think schooldays are/were the best days of your life?			

2 Listen to Gavin and Jenny, who are English, answering the questions. Put a tick (✓) by the ones they say *yes* to.

3 Work in pairs and check your answers to 2. Can you remember what Gavin and Jenny said in detail?

Now listen again and check.

4 Work in pairs. Ask your partner the questions in the chart.

5 Work in groups of three or four.

Student A, B, C: Student D has chosen an adverb from the vocabulary box. Ask him/her to:

– say something
– sing
– perform an action
– read something from *Reward* Elementary

He/she will perform the action in the manner of the adverb. Try to guess the adverb.

Student D: Choose an adverb from the vocabulary box. Don't tell the others what it is. The students in your group will ask you to do something in the manner of the adverb you have chosen. They must guess the adverb you have chosen.

Change round when you're ready.

Progress check 31–35

VOCABULARY

1 A collocation is two or more words which often go together.

fast car   high mountain   busy street   have dinner

Here are some adjectives from Lessons 31 to 35.

cold   difficult   low   old   expensive

Think of nouns which often go with the adjectives. You can use your dictionary.

cold day

2 There may be many places outside your classroom where you can see and listen to English, and build your vocabulary. In which of the following places can you see or listen to English words?

– food labels
– the airport
– notices and signs
– instructions (for electrical goods)
– the station
– the radio
– travel documents (tickets etc)
– the television
– newspapers

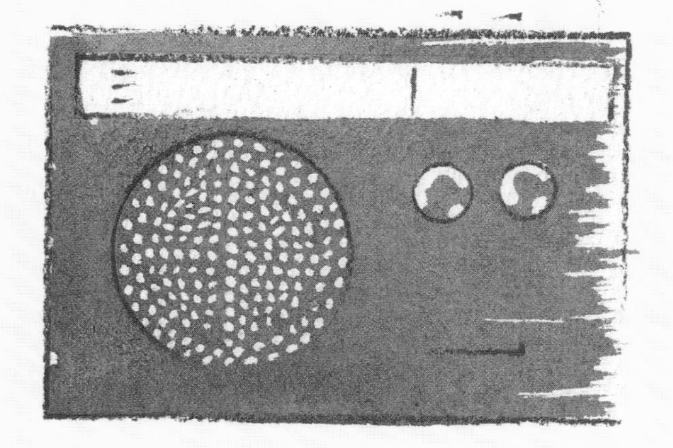

GRAMMAR

1 Complete these sentences from Lesson 31 with *must* or *mustn't.*

1 In parts of Africa you _____ ask if you want to take a photograph of someone.
2 In Saudi Arabia women _____ cover their heads in public.
3 In China you _____ kiss in public.
4 In Japan you _____ look people in the eye.
5 In Taiwan you _____ give a gift with both hands.
6 In Britain you _____ point at people.

2 Write sentences saying why you:

1 use a blanket
2 go to the swimming pool
3 go to the supermarket
4 use a match
5 use a bottle opener
6 use the telephone

1 You use a blanket to keep warm.

3 Write the adverbs which come from these adjectives.

easy   good   careful   fast   hard   polite   quiet   rude

4 Complete the sentences with the adjective in brackets or the adverb which comes from it.

1 She speaks English very _____. (good)
2 He drives extremely _____. (fast)
3 He's extremely _____ to people. (polite)
4 Don't make so much noise. Please be _____. (quiet)
5 Hungarian is a _____ language to learn. (hard)
6 He passed his First Certificate exam _____.(easy)

Progress check 31 - 35

GENERAL COMMENTS

You can work through this Progress check in the order shown, or concentrate on areas which have caused difficulty in Lessons 31 - 35. You can also let the students choose the activities they would like or feel the need to do.

VOCABULARY

1 Aim: to focus on collocations.

- There has already been quite a lot of work on collocations in *Reward* Elementary, but the students have not yet used the specific terms.
- Ask the students to think of other words which go with the adjectives.

Possible answers
cold day, weather
difficult exam, time
low land, bridge
old lady, man
expensive holiday, car

2 Aim: to focus on English outside the classroom.

- Encourage the students to think of all the different places outside the classroom. The list in this activity isn't exhaustive, but it's designed to demonstrate to the students that even in small towns, there may be opportunities to come across examples of English.

GRAMMAR

1 Aim: to revise *must* and *mustn't*.

Answers
1 must 2 must 3 mustn't
4 mustn't 5 must 6 mustn't

2 Aim: to revise the infinitive of purpose.

Answers
1 to keep warm
2 to go swimming
3 to go shopping
4 to light something
5 to open a bottle
6 to speak to someone at a distance

3 Aim: to revise the formation of adverbs.

Answers
easily, well, carefully, fast, hard, politely, quietly, rudely

4 Aim: to revise the distinction between adjectives and adverbs.

Answers
1 well 2 fast 3 polite 4 quiet 5 hard 6 easily

SOUNDS

1 Aim: to focus on /eə/ and /ɪə/.

- Remind the students once again that the relationship between spelling and pronunciation in English is not as close as in other languages.

- Ask the students to group the words according to the diphthongs.

Answers
/eə/: beer, near, hear, we're, year
/ɪə/: stair, hair, chair, air

- Play the tape and ask the students to say the words aloud.

2 Aim: to focus on /ɔː/ and /əʊ/.

- These are two very difficult phonemes for most students, so don't expect them to pronounce them correctly.

Answers
/ɔː/: so, Jo, go, low, toe
/əʊ/: tore, war, floor, law, sore, sport

- Play the tape. As they listen, ask the students to say the words aloud.

3 Aim: to focus on stressed words and their effect on meaning.

- Explain that the speaker will stress the word he or she thinks is important, and this will influence the meaning of the whole sentence.

- Play the tape and ask the students simply to read and follow the tapescript.

- You may like to ask the students to practise the exchanges in pairs.

READING AND WRITING

1 Aim: to prepare for reading.

- Ask the students to look at the pictures and predict what happens in the story.

- Ask the students to read the story and check the predictions.

2 Aim: to focus on word order.

- Ask the students to cross out unnecessary words, i.e. words which can be left out without creating a syntactical error or without changing the meaning.

- You may need to explain the meaning of any difficult words.

3 Aim: to check the answer to 2.

- Ask the students to work in pairs and to check their answers.

- Find out how many words the students have crossed out.

Possible answers
A young man went into a local bank, went up to the woman cashier and gave her a note and a plastic bag. The note said, Put all your money into this bag, please. The middle-aged cashier was very frightened so she gave him all the money. He put it in his bag and ran out of the front door. When he got back home the city police were there. His note was on an old, white envelope and on the envelope was his home address.

4 Aim: to focus on unnecessary words.

- You will find that most passages will contain some 'unnecessary' words. If, by any unlucky chance, the passage chosen does not contain unnecessary words, even the process of looking for them will be a worthwhile activity.

- You may like to ask your students to do this activity for homework.

SOUNDS

1 Group the words with the same vowel sound.

beer stair hair near hear chair air we're year

Listen and check. As you listen, say the words aloud.

2 Say these words aloud. Is the underlined sound /əʊ/ or /ɔː/?

go so Jo law tore war floor low sore toe sport

Listen and check. As you listen, say these words aloud.

3 Listen to how you can change the stressed word in a question and get a different answer.

1 a Can you **speak** Spanish?
No, but I can **write** it.
b Can you speak **Spanish**?
No, but I can speak **Italian**.

2 a Did you stay with friends in **Paris**?
No, I stayed with friends in **Rome**.
b Did you stay with **friends** in Paris?
No, I stayed in a **hotel**.

3 a Have you got this **dress** in another colour?
No, only the **jeans**.
b Have you got this dress in another **colour**?
No, we've only got it in **red**.

READING AND WRITING

1 Look at the pictures below. Can you guess what happens in the story. Now read the story and see if you guessed correctly.

2 Read the story again and cross out any words which aren't 'necessary'. You cannot cross out two or more words together.

A young man went into a local bank, went up to the woman cashier and gave her a note and a plastic bag. The note said, 'Put all your money into this bag, please.' The middle-aged cashier was very frightened so she gave him all the money. He put it in his bag and ran out of the front door. When he got back home the city police were there. His note was on an old, white envelope and on the envelope was his home address.

3 Work in pairs and check your answers to 2.

4 Work in pairs and choose a short passage from *Reward* Elementary.

Working alone, count the number of 'unnecessary' words in the passage.

Tell each other how many words you've found. Who has found the most?

36 I'll go by train

Future simple (1): *(will)* for decisions

VOCABULARY AND SOUNDS

1 Look at the words in the box. Put them under two headings: *train* and *plane*. Use a dictionary, if necessary.

departure lounge passport control baggage reclaim check-in arrival hall platform boarding pass business class first class cheap day return return single departure gate ticket office tourist class

2 Which is the odd word out?

1 departure lounge business class check-in
2 platform ticket office departure gate
3 tourist class first class arrival hall
4 ticket office passport control boarding pass

3 Listen to these two-word nouns. Underline the stressed word.

departure lounge passport control business class ticket office arrival hall

Now say the words aloud.

LISTENING

1 Look at this conversation. Where does it take place?

A Can I help you?
B Yes, I'd like a ticket to London.
A When do you want to travel? It's cheaper after 9.15.
B I'll travel after 9.15.
A Single or return?
B I'll have a cheap day return ticket, please.
A That'll be thirteen pounds exactly. How would you like to pay?
B Do you accept credit cards?
A I'm afraid not.
B Well, I'll pay cash, then. Will there be refreshments on the train?
A No, I'm afraid there won't.
B Can I have a ticket for the car park as well.
A That'll be fifteen thirty in all.
B Thank you.

2 Listen and underline anything which is different from what you hear.

36

GENERAL COMMENTS

Future simple

This is the first of two lessons in which the future simple is presented. At this level, the students are not expected to be able to use this tense expertly. They simply need to know how it is formed, which is relatively straightforward, and its two principle uses: to talk about a decision made at the time of speaking (Lesson 36) and to make predictions (Lesson 37). Further work on the future simple and its uses is provided in later levels of the *Reward* series.

VOCABULARY AND SOUNDS

1 Aim: to present the words in the box.

- Ask the students to look at the words, and to check they understand them. Ask other students to explain the meaning of the words, if necessary.
- Write on the board *train* and *plane*. Ask the students to come up in turn and to write a word in the correct column. Some words can go in both columns.

Answers
train: platform, first class, cheap day return, return, single, ticket office
plane: departure lounge, passport control, baggage reclaim, check-in, arrival hall, boarding pass, business class, first class, return, single, departure gate, tourist class

2 Aim: to check comprehension of the new vocabulary.

- Ask the students to do this activity on their own and then to check their answers in pairs.

Answers
1 business class
2 departure gate
3 arrival hall
4 boarding pass

- There may be other odd words out. Ask your students to explain their answers.

3 Aim: to practise pronouncing two-word nouns.

- Play the tape and ask the students to underline the stressed word.

Answers
departure lounge, passport control, business class, ticket office, arrival hall

- Ask the students to say the words aloud.

LISTENING

1 Aim: to prepare for listening; to practise reading for main ideas; to present the future simple.

- Ask the students to read the conversation and to say where it takes place.

Answer
At a railway ticket office.

2 Aim: to practise listening for specific information.

- Play the tape and ask the students to listen and underline anything which is different from what they hear.

Answers
A Can I help you?
B Yes, I'd like a ticket to Birmingham.
A When do you want to travel? It's cheaper after 9. 30.
B I'll travel after 9. 30.
A Single or return?
B I'll have a single ticket, please.
A That'll be thirty pounds exactly. How would you like to pay?
B Do you accept credit cards?
A I'm afraid not.
B Well, I'll pay cash, then. Will there be refreshments on the train?
A Yes, there will.
B Can I have a ticket for the car park as well?
A That'll be thirty-two pounds in all.
B Thank you.

- Ask the students to check their answers in pairs.

GRAMMAR

1 Aim: to practise using the future simple.

• Ask the students to read the information in the grammar box, and then to do the exercises.

• Draw the students' attention to the future simple in the conversation and explain that it's used when you talk about a decision at the moment of speaking.

• Divide the students into pairs of Student A and Student B. Ask each student to change some details in the conversation.

2 Aim: to practise speaking.

• Ask the students to act out the conversation when they're ready.

• You may like to ask the students to act out their conversations in front of the class.

READING AND SPEAKING

1 Aim: to practise reading for main ideas.

• Ask the students if they would like to go to Africa. Ask them what they would like to do or see there. Find out if anyone has already been there.

• Ask the students to read the travel brochure and to follow the route on the map.

Answers
Cape Town to Kimberley
Kimberley to Johannesburg and Pretoria
Pretoria to the Victoria Falls
Livingstone to Windhoek
Windhoek to London

2 Aim: to practise reading for specific information; to check comprehension.

• This activity provides an opportunity for a closer look at the passage. You may want to explain a few words, but by now, the students should be able to accept that they won't be able to understand every single word.

Answers
1 Five days.
2 Three days.
3 Only for meals in the hotel.
4 By plane.
5 Only in the hotels.
6 Yes.
7 Travel insurance, visa, airport taxes, tips.
8 Yes.

3 Aim: to practise reading and speaking.

• The Communication activities contain extra information for a role play between tourists and tour organisers.

• Ask the students to read the relevant Communication activity and to act out the role play.

• You may like to ask the students to write out their role plays and learn them for homework. At the start of the next lesson you can ask some of them to perform their role plays to the rest of the class.

GRAMMAR

Future simple (1): *(will)* for decisions

You form the future simple with *will* or *won't* + infinitive.

I		
you	*'ll (will)*	
he/she/it	*won't (will not)*	*go by train.*
we		
they		

You use *will* when you make a decision at the time of speaking.

*I'**ll** have a return ticket.* *I'**ll** pay cash.*

1 Work in pairs.

Student A: You're going to act out the conversation in *Listening* activity 1. You're in the ticket office. Change some of the details in the conversation.

Student B: You're going to act out the conversation in *Listening* activity 1. You're the passenger. Change some of the details about the ticket you want to buy.

2 Act out the conversation when you're ready.

READING AND SPEAKING

1 Read the travel brochure and follow the route on the map.

2 Imagine you want to go on the tour described in the travel brochure. Look at the brochure and answer the questions.

1 How long will you spend on the train?
2 How long will you spend in Cape Town?
3 Will you have to pay extra for meals?
4 How will you get from Livingstone to Windhoek?
5 Will you be able to have a single room?
6 Will you be able to go to a game reserve?
7 Will there be anything extra to pay?
8 Will you be able to stay an extra day?

3 Work in groups of three.

Student A: Turn to Communication activity 19 on page 104.
Student B: Turn to Communication activity 9 on page 100.
Student C: Turn to Communication activity 11 on page 101.

TRAVEL IN STYLE – FROM THE CAPE TO VICTORIA FALLS

Visiting Cape Town, Kimberley, Pretoria and the Victoria Falls.
Including a trip on the Pride of Africa – probably the finest train in the world!

FOR ONLY £2595 per person

ITINERARY

Day 1 Leave London and fly overnight to Cape Town.

Day 2 – 4 Sightseeing in Cape Town.

Day 5 – 7 Join the train and travel overnight to Kimberley, and on to Johannesburg and Pretoria.

Day 8 Leave Pretoria by train.

Day 9 Travel all day across Zimbabwe, and on to the Victoria Falls.

Day 10 – 12 Leave the train. Sightseeing around the Victoria Falls and a visit to Chobe game reserve.

Day 13 Take the plane from Livingstone to Windhoek. Connect with flight to London.

Day 14 Arrive in London.

Accommodation in a double sleeping cabin on the train and in a double room in the Cape Town and Chobe Game Reserve. Single beds available in the hotels. Extra night's accommodation in Windhoek available.

Facilities in hotel: swimming pool, restaurant, bar, tennis.

Price per person includes air travel, all meals on the train, bed and breakfast in the hotels, transfer to and from the airport. Not included: travel insurance, visa, airport taxes, tips.

What will it be like in the future?

Future simple (2): *(will)* for predictions

VOCABULARY AND LISTENING

1 Match the words in the box to the symbols for weather.

fog cloud sun rain wind snow

2 Which of these adjectives can you use to describe today's weather?

cold cool dry foggy hot rainy sunny warm wet windy snowy

It's very hot today.

3 Look at the photos and say what the weather is like.

4 Work in pairs. Look at the newspaper weather report below and listen to the radio weather forecast. Underline any information which is different from what you hear.

Worldwide forecast for midday tomorrow

Athens	c	12
Bangkok	c	30
Cairo	s	16
Geneva	c	4
Hong Kong	c	17
Istanbul	r	7
Kuala Lumpur	c	30
Lisbon	c	11
Madrid	r	7
Moscow	sn	−10
New York	s	0
Paris	sn	6
Prague	sn	−2
Rio	c	29
Rome	r	9
Tokyo	c	4
Warsaw	c	−8

GRAMMAR

Future simple (2): *(will)* for predictions

You form the future simple with *will* + infinitive. You use the future simple to make predictions.

*It'**ll be** sunny in New York tomorrow.*
(= It will be sunny in New York tomorrow.)
*It **won't be** rainy.* (= It will not be rainy.)
Will it be snowy in New York? — *Yes, it will.*
Will there be rain in Geneva? — *No, there won't.*

1 Work in pairs. Correct your answers to *Vocabulary and listening* activity 4.

In Geneva, it will be cloudy and ten degrees.

37

GENERAL COMMENTS

The weather

This lesson covers some very simple vocabulary to describe the weather. You may want to add extra words to describe the weather in the place where you're teaching at the moment.

In our research for *Reward*, we have found that despite the importance of the topic, many teachers and students are less willing than perhaps they were to discuss environmental matters. This is possibly because it is a subject which is covered extensively in newspapers and other media. It may also be that it is a subject which is not discussed a great deal outside the classroom. You may want to be vigilant of spending too much time on a topic which, while it constitutes an important lexical field, may not maintain the interest of the students.

VOCABULARY AND LISTENING

1 Aim: to present the words in the vocabulary box.

- Teach the meaning of the words by asking the students to match them with the weather symbols.
- You may need to explain some of the symbols if they represent weather with which the students are unfamiliar.

Answers
fog, snow, wind, cloud, sun, rain

2 Aim: to present the words in the vocabulary box.

- You may need to translate or illustrate these words in some way if your students do not understand them at first. Weather stereotypes such as *foggy London, snowy Austria*, may help you, especially if you come from a country which doesn't have these extremes of weather.

3 Aim: to practise saying what the weather is like.

- Ask the students to say what the weather is like in the photos.
- Ask the students to say what the weather is like today, and what it was like yesterday.
- Try to match some of the words with different months or seasons of the year. You may also like to bring some magazine photos of different outdoor scenes from different countries to illustrate the words. If you can collect enough to illustrate each word, hand out the photos and ask the students to go round saying *What's the weather like?* and eliciting a reply which is appropriate to the weather in the photo.

4 Aim: to practise listening for specific information.

- Play the weather forecast and ask the students to underline anything which is different from what they hear.

Answers
And here's the weather report for the rest of the world. Athens, cloudy twelve degrees. Bangkok, cloudy thirty degrees. Cairo, sunny sixteen degrees. Geneva, cloudy ten degrees. Hong Kong, cloudy twenty degrees. Istanbul, rainy seven degrees. Kuala Lumpur, sunny thirty-five degrees. Lisbon, cloudy eleven degrees. Madrid, rainy seven degrees. Moscow, snowy minus ten degrees. New York, sunny zero degress. Paris, snowy minus six. Prague, sunny minus two.Rio, cloudy minus twenty-nine. Rome, rainy nine degrees. Tokyo, snowy minus four degrees. Warsaw, cloudy minus eight degrees.

- You may like to spend some time asking the students to say what country the different cities are in.

GRAMMAR

1 Aim: to focus on the form of the future simple for predictions.

- Ask the students to read the information in the grammar box and then to do the exercises.
- Remind the students that the two meanings of the future simple presented in *Reward* Elementary are decisions made at the moment of speaking (Lesson 36) and predictions (Lesson 37).
- This is a fairly mechanical, drill-like activity, but at least it will give the students plenty of opportunity of using the future simple and the words for describing the weather. You may like to check their answers with the whole class.

2 Aim: to practise using comparative forms of the adjectives in the vocabulary box; to practise using the future simple for predictions.

● Make sure the students refer to their corrected versions of the weather forecast.

● Ask the students to make comparisons between the two towns.

Possible answers
1 It'll be hotter in Cairo than in Tokyo.
2 It'll be hotter in Athens than in Rome.
3 It'll be colder in Moscow than in Warsaw.
4 It'll be hotter in Bangkok than in Lisbon.
5 It'll be colder in Prague than in Madrid.
6 It'll be colder in Istanbul than in Athens.

3 Aim: to practise the future simple for predictions.

● Ask the students to write a short forecast for tomorrow.

● Find out if everyone has made similar predictions.

● Tell the students that in Britain and Northern Europe the weather is very unpredictable and people say you can experience all four seasons in one day, all the year round. Is their country like that?

READING AND LISTENING

1 Aim: to practise using the target structures; to prepare for reading.

● Ask the students to look at the predictions about the weather in twenty-five years' time and to decide if they're true for their country.

● Check the students' predictions with the whole class.

2 Aim: to practise reading for main ideas and for specific information.

● The predictions in 1 constitute a framework for the main ideas of the passage. Although the students will read the passage for specific information, they will also get a clear picture of the passage's main ideas as a whole.

Answers

It'll be colder.	False
The sea level will be lower.	False
It'll be windier.	True
It'll be wetter.	True, in the north
There'll be more snow.	False

3 Aim: to prepare for listening; to practise speaking.

● Ask the students to discuss these predictions, either in small groups or with the whole class.

4 Aim: to practise listening for main ideas.

● The students may not understand every word of what the scientist says, but they will understand enough to perform the activity.

● Play the tape and ask the students to put T if the statements in activity 3 are true and F if they are false.

Answers
1T 2T 3F 4F 5T 6T 7F

5 Aim: to check comprehension; to practise speaking.

● Ask the students to check their answers. Ask them if they can remember what the scientist said in detail.

● Play the tape again.

6 Aim: to practise speaking; to practise using the target structures.

● Ask the students to continue the discussion in groups of two or three. Ask them to make predictions about the following: *population, medical research, the economy, politics, transport*

7 Aim: to practise speaking.

● Ask the students to tell other students about their predictions. How many groups made similar predictions?

2 `Look at the weather report again. Make comparisons between the weather in these cities.

1 Cairo – Tokyo
2 Athens – Rome
3 Warsaw – Moscow
4 Bangkok – Lisbon
5 Prague – Madrid
6 Istanbul – Athens

1 It'll be hotter in Cairo than in Tokyo.

3 Write a short forecast for tomorrow's weather in your town.

Tomorrow it'll be sunny with some clouds.

READING AND LISTENING

1 Look at these predictions about the weather. Do you think they are true for your country?

In twenty-five years:
– it'll be colder
– the sea level will be lower
– it'll be windier
– it'll be wetter
– there'll be more snow

2 Read *The temperature's rising* and find out if the predictions in 1 are true for Britain.

The temperature's rising

A government report says that in the next twenty-five years, Britain will get warmer and have higher sea levels. The weather will become more Mediterranean, and tourism will grow, but the Scottish ski industry will disappear because there will be little or no snow, and there'll be stronger winds. In the South there will be more sun, enough to produce wine, and in the North there will be more rain. It will be good for farmers, as crops will grow more quickly, and cattle and sheep will have warmer and wetter land in Scotland and northern England. But the higher sea level means that many towns, including London, will disappear under water.

People will only use heating in their homes for two or three months of the year, but they'll pay more for water. Snow at Christmas will become very rare. More people will die in the hotter summers, but the winters will be warmer as well.

3 Here are some more worldwide predictions. Which do you think will be true?

1 Temperatures will rise by two to six degrees Celsius in twenty-five years' time.
2 Ice at the North and South Pole will melt.
3 Whole countries will disappear underwater.
4 There won't be enough fresh water for everyone.
5 Fresh water will cost more.
6 Factory goods will cost more to produce.
7 The world economy will get worse.

4 Listen to a scientist talking about the predictions in 3. Put T if the predictions are true and F if they are false.

5 Work in pairs and check your answers to 4. Can you remember any details?

Listen again and check.

6 Work in groups of two or three. Make other predictions about the future.

There'll be more people in the world.

7 Tell other people about your predictions. Do they agree with you?

38 Hamlet was written by Shakespeare

Active and passive

SPEAKING AND VOCABULARY

1 Work in pairs. Are these sentences true or false?

Apple makes computers.
Lemons grow on trees.
Gustav Eiffel built the Eiffel Tower.
Marconi invented the radio.
Beethoven composed the Moonlight Sonata.
Leonardo da Vinci painted the Mona Lisa *(La Joconde)*.
Shakespeare wrote Hamlet.
Fleming discovered penicillin.

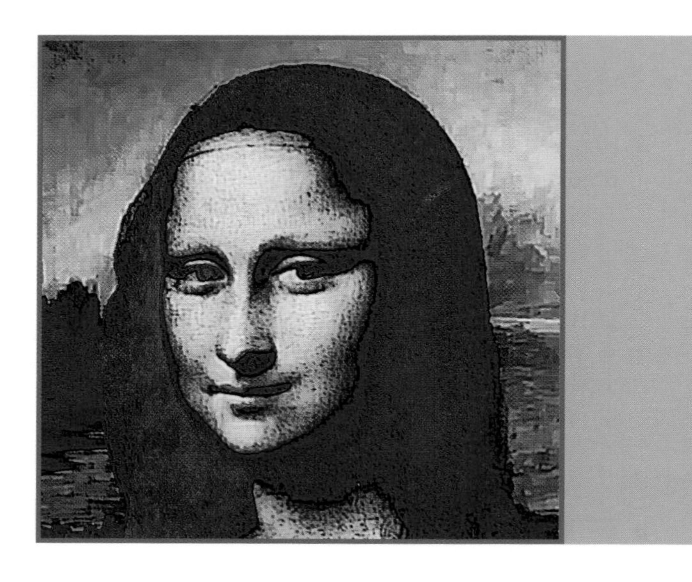

2 Complete these sentences with verbs from the box. You may have to change the tense.

make grow build invent compose paint write discover

1 Botticelli _____ *La Primavera*.
2 They _____ oranges in Spain.
3 Columbus _____ America.
4 Shah Jehan _____ the Taj Mahal.
5 Homer _____ the Odyssey.
6 They _____ Fiat cars in Italy.
7 The Chinese _____ gunpowder.
8 Tchaikovsky _____ the 1812 Symphony.

READING

1 Read *The round-the-world quiz* and choose the correct answer.

The round-the-world quiz

1 Coffee is grown in ...
a Brazil b England c Sweden

2 Daewoo cars are made in ...
a Switzerland b Thailand c Korea

3 Sony computers are made in ...
a Japan b the USA c Germany

4 Tea is grown in ...
a India b France c Canada

5 Tobacco is grown in ...
a Norway b Iceland c the USA

6 Benetton clothes are made in ...
a Italy b France c Malaysia

7 Roquefort cheese is made in ...
a Germany b Thailand c France

8 The atom bomb was invented by ...
a the Japanese b the Americans c the Chinese

9 *Guernica* was painted by ...
a Picasso b Turner c Monet

10 The West Indies were discovered by
a Scott of the Antarctic b Christopher Columbus
c Marco Polo

11 The telephone was invented by ...
a Bell b Marconi c Baird

12 *Romeo and Juliet* was written by ...
a Ibsen b Shakespeare c Primo Levi

13 The Blue Mosque in Istanbul was built by
a Sultan Ahmet I b Ataturk
c Suleyman the Magnificent

14 *Yesterday* was composed by ...
a Paul McCartney b John Lennon c Mick Jagger

15 The Pyramids were built by ...
a the Pharaohs b the Sultans c the Council

2 Work in pairs. Check your answers to the quiz.

38

GENERAL COMMENTS

Active and passive

The passive voice is introduced for the first time in this lesson. The form of the passive is perfectly accessible. You use the various tenses of the verb *to be* + past participle. However, the full range of uses are not so easy to explain. This lesson restricts itself to the way the passive allows you to shift the focus from the object of an active verb to the subject. The other uses are covered more fully in later levels of *Reward*.

The round-the-world quiz

The quiz in this lesson assumes that the students have a certain level of general knowledge. The questions are as simple as possible and drawn from a wide geographical and cultural background. However, it doesn't matter if the students have difficulty in answering them. The primary aim of the lesson is to introduce the passive. Testing the students' general knowledge is of secondary importance.

SPEAKING AND VOCABULARY

1 Aim: to practise speaking.

- As a warm-up for the rest of the lesson, which involves a general knowledge quiz, ask the students to read the statements and say if they are true.

- You may like to check the students' answers with the whole class, or if you think they may have difficulty getting the correct answers, you may decide to give them the answers without finding out who got them right.

Answers
The sentences are all meant to be true, although some of the claims have been disputed (Shakespeare, Marconi).

2 Aim: to present the words in the vocabulary box.

- These words are among the most common ones to be used in the passive. They will therefore be useful later in the lesson.

- Remind the students that the sentences in 1 were true. They should be able to transfer the meaning of the new vocabulary to complete the sentences in this activity.

Answers
1 painted 2 grow 3 discovered 4 built
5 wrote 6 make 7 invented 8 composed

READING

1 Aim: to practise reading for specific information.

- Ask the students to read *The round-the-world quiz* and to tick the correct answers.

2 Aim: to practise speaking; to check comprehension.

- Ask the students to check their answers in pairs.

- You shouldn't give them the correct answers at this stage, because this will detract from the value of the *Listening* activity.

GRAMMAR

1 Aim: to focus on the form of the passive.

● Ask the students to read the information in the grammar box and then to do the exercises.

● Ask the students to check how the passive is formed. Tell them that there are other tenses which are used in the passive, but in this lesson, only the present and the past passive are covered.

Answers
Sentences 1 – 7: present passive
Sentences 8 – 15: past passive

2 Aim: to focus on the form of the passive.

● This transformation activity from the passive back to the active should give the students further help in understanding how the passive is formed. Make sure they use *they* in the present active sentences.

● You may like to check the answers with the whole class.

Answers
They grow coffee in Brazil.
They make Daewoo cars in Korea.
They make Sony computers in Japan.
They grow tea in India.
They grow tobacco in the USA.
They make Benetton clothes in Italy.
They make Roquefort cheese in France.
The Americans invented the Atom bomb.
Picasso painted *Guernica*.
Christopher Columbus discovered the West Indies.
Bell invented the telephone.
Shakespeare wrote *Romeo and Juliet*.
Sultan Ahmet I built the Blue Mosque.
Paul McCartney composed *Yesterday*.
The Pharaohs built the Pyramids.

3 Aim: to practise writing the passive.

● By now, the formation of the passive should be causing few problems. Ask the students to rewrite the sentences in *Speaking and vocabulary* activity 2 in the passive.

● Once again, you may like to correct this activity with the whole class.

Answers
1 *La Primavera* was painted by Botticelli.
2 Oranges are grown in Spain.
3 America was discovered by Columbus.
4 The Taj Mahal was built by Shah Jehan.
5 The *Odyssey* was written by Homer.
6 Fiat cars are built in Italy.
7 Gunpowder was invented by the Chinese.
8 The *1812 Overture* was composed by Tchaikovsky.

LISTENING

1 Aim: to practise listening for main ideas.

● Having done the quiz, the students should be well-prepared for this listening activity. Remind them that they don't have to understand everything that Frank and Sally say, only their answers.

● Play the tape.

Correct answers
1 Brazil 2 Korea 3 Japan 4 India 5 the USA
6 Italy 7 France 8 the Americans 9 Picasso
10 Christopher Columbus 11 Bell 12 Shakespeare
13 Sultan Ahmet I 14 Paul McCartney
15 The Pharaohs

Frank and Sally scored 15.

2 Aim: to check the students' answers to *Reading* activity 2.

● Ask the students to check their own score of correct answers.

WRITING

1 Aim: to practise writing.

● Ask the students to work in pairs and to write a few quiz questions. Ask them to think of one correct and two incorrect answers to each question.

● You may like to ask them to do this activity for homework.

2 Aim: to practise reading.

● Collect the quizzes and give them out to new pairs. Ask the new pair to do the quiz.

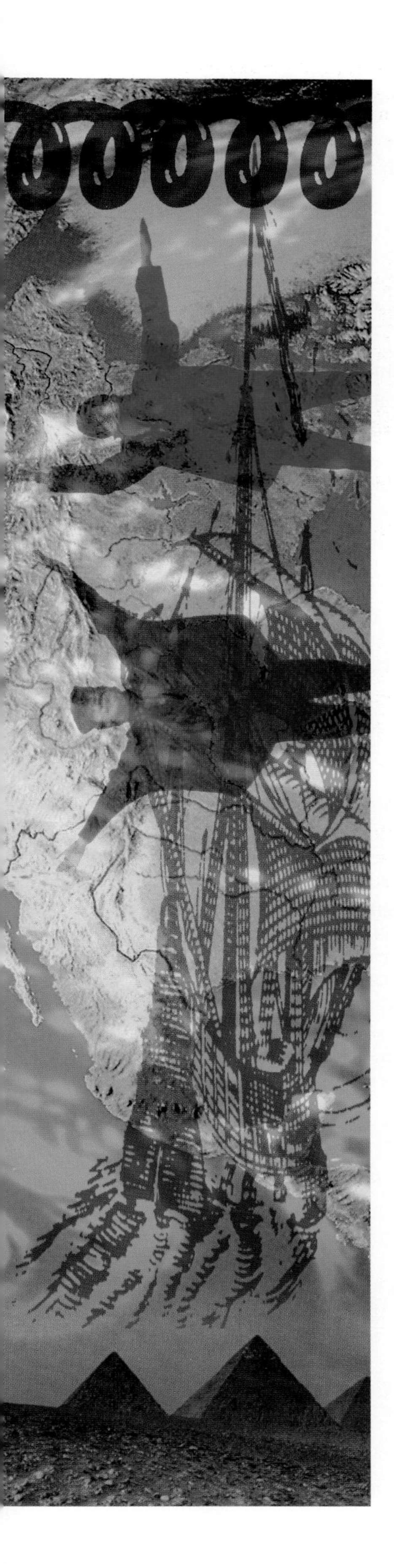

GRAMMAR

Active and passive

Active	**Passive**
They ***grow*** *coffee in Brazil.*	*Coffee* ***is grown*** *in Brazil.*
The Americans ***invented*** *the atom bomb.*	*The atom bomb* ***was invented*** *by the Americans.*

You form the passive with the verb *to be* + past participle. You use *by* to say who or what is responsible for an action.

Present simple	**Past simple**
Tea ***is grown*** *in India.*	*Guernica was painted* ***by*** *Picasso.*
Daewoo cars ***are made*** *in Korea.*	*The Pyramids were built* ***by*** *the Pharaohs.*

1 Look at the quiz again. Find examples of the passive. Are they in the present or past simple?

Sony computers are made in Japan.

2 Choose six sentences from the quiz and rewrite them.

Bell invented the telephone. They grow tobacco in the USA.

3 Rewrite the sentences in *Speaking and vocabulary* activity 2 in the passive.

La Primavera was painted by Botticelli.

LISTENING

1 Listen to Frank and Sally doing *The round-the-world quiz*. Tick (✓) the correct answers. How many do they score?

2 How many correct answers did you get?

WRITING AND SPEAKING

1 Work in pairs. Write a quiz about your country.

The Chapel Royal in the Wat Phrae Kaew was built by ....

a King Rama 1   *b King Rama IV*   *c King Rama V*

Make sure you include one correct ending and two incorrect ones.

2 Work with another pair. Do each other's quizzes.

39 *She said it wasn't far*

Reported speech: statements

LISTENING AND READING

1 Decide where these sentences go in the conversation.

a ... it leaves at nine o'clock in the evening.
b And where's the next hostel?
c Where are you walking to?
d Yes, it's two kilometres.
e We'll stay for just one night.

CHRIS Good afternoon.

RECEPTIONIST Good afternoon. Can I help you?

CHRIS Have you got any beds for tonight?

RECEPTIONIST Yes, I think so. Sorry, but I've just started work at the hostel. How long would you like to stay?

CHRIS (1) _____

RECEPTIONIST Yes, that's OK.

TONY Great!

RECEPTIONIST How old are you?

TONY We're both sixteen.

RECEPTIONIST One night's stay costs £6.50 each.

CHRIS Is it far from the hostel to the centre of Canbury?

RECEPTIONIST (2) _____ It takes an hour on foot.

TONY Is there a bus service?

RECEPTIONIST I think so. It takes about fifteen minutes. There's a bus every hour.

TONY When does the last bus leave the city centre?

RECEPTIONIST I think (3) _____ There's not much to do in the evening.

CHRIS We're very tired. We need an early night. What time does the hostel close in the morning?

RECEPTIONIST Er, at eleven am. (4) _____

CHRIS We're going to Oxton. Are you serving dinner tonight?

RECEPTIONIST Yes, we're serving dinner until eight o'clock. And breakfast starts at seven-thirty.

TONY (5) _____

RECEPTIONIST I'm not sure. I think it's Kingscombe, which is about ten kilometres away. I started work last Monday so I'm very new here.

2 Listen and check.

3 The receptionist gives Chris and Tony some wrong information. Read the brochure and underline the wrong information in the conversation.

Canbury Youth Hostel,
Jackson Lane, Canbury
Tel 01789 3445 Fax 01789 3446

Facilities
112 beds
Open 24 hours, all year
Car park
No family rooms
Games room
Washing machines
No smoking hostel
Camping

Charges
Under 18 £6.15 Adult £9.10

Meal times
Breakfast 7am
Dinner 6pm - 7pm

Travel Information
City centre three kilometres
Bus service to city centre No.14
(takes ten minutes) Last bus 8pm

Next hostels
Charlestown 8 kilometres
Kingscombe 15 kilometres

39

GENERAL COMMENTS

Reported speech

In this lesson, a very brief presentation of reported speech is given, in which only the tense shift from present simple to past simple is given. Students will be given more extensive coverage of reported speech, reported questions and reporting verbs in *Reward* Intermediate and Upper intermediate.

Youth hostels

You may like to explain that youth hostels are cheap forms of accommodation, both for teenagers and for families. In Britain, they are mostly within walking distance of each other and many teenagers go on 'youth hostelling' walking tours of the country with their friends.

LISTENING AND READING

1 Aim: to prepare for listening; to practise understanding text organisation.

- Ask the students to read the conversation and to decide where the sentences go.
- Ask the students to check their answers with another student.

2 Aim: to practise listening for specific information.

- Having read the conversation the students will be ready for this listening activity. Play the tape and ask them to listen and check their answers to 1.

Answers
1e 2d 3a 4c 5b

3 Aim: to practise reading for specific information; to prepare the presentation of reported speech.

- This activity involves some close reading of the brochure and the conversation.
- Ask the students to read the brochure and the conversation and to underline any information which is different.

Answers

CHRIS	Good afternoon.
RECEPTIONIST	Good afternoon. Can I help you?
CHRIS	Have you got any beds for tonight?
RECEPTIONIST	Yes, I think so. Sorry, but I've just started work at the hostel. How long would you like to stay?
CHRIS	We'll stay for just one night.
RECEPTIONIST	Yes, that's OK.
TONY	Great!
RECEPTIONIST	How old are you?
TONY	We're both sixteen.
RECEPTIONIST	One night's stay costs £6.50 each.
CHRIS	Is it far from the hostel to the centre of Canbury?
RECEPTIONIST	Yes, it's two kilometres. It takes an hour on foot.
TONY	Is there a bus service?
RECEPTIONIST	I think so. It takes about fifteen minutes. There's a bus every hour.
TONY	When does the last bus leave the city centre?
RECEPTIONIST	I think it leaves at nine o'clock in the evening. There's not much to do in the evening.
CHRIS	We're very tired. We need an early night. What time does the hostel close in the morning?
	Er, at eleven a.m. Where are you walking to?
CHRIS	We're going to Oxton. Are you serving dinner tonight?
RECEPTIONIST	Yes, we're serving dinner until eight o'clock. And breakfast starts at seven-thirty.
TONY	And where's the next hostel?
RECEPTIONIST	I'm not sure. I think it's Kingscombe, which is about ten kilometres away. I started work last Monday so I'm very new here.

4 Aim: to present reported speech.

- This exchange focuses on the tense shift which reported speech requires.
- Play the tape. Ask the students to put the parts of the conversation below in the order they hear them. Don't spend too long on this as the text organisation aspect of the activity is less important than the presentation of reported speech.

Answers

1	CHRIS	It's very strange. She said one night's stay cost £6.50, but it costs £6.15.
2	TONY	Yes, and she said it was two kilometres to the city centre.
3	CHRIS	But, in fact, it's three kilometres.
4	TONY	And she said the last bus left at nine o'clock. But it leaves at eight o'clock.

GRAMMAR

1 Aim: to focus on the tense shift and pronoun change in reported speech.

- Ask the students to read the information in the grammar box and then to do the exercises.

- Do this activity orally with the whole class. Ask the students to look back at the transcript of the conversation.

Answers

1 'We're both sixteen,' said Tony.
2 'It takes an hour on foot,' said the receptionist.
3 'There's a bus every hour,' she said.
4 'There isn't much to do in the evening,' she said.
5 'We're very tired,' said Chris.
6 'We need an early night,' he said.

2 Aim: to focus on the tense shift in reported speech; to listen to the conversation in *Listening and reading* 4; to provide material for writing the letter of complaint.

- This activity performs a variety of functions. Ask the students to look at the rest of the conversation and to complete it.

- You may like to ask the students to check this in pairs.

- Ask several students to perform the conversation to the rest of the class.

Answers

CHRIS And she said the bus took fifteen minutes. But in fact, it takes ten minutes.
TONY And she said the hostel closed at eleven am, but it's open all day.
CHRIS It seems that they serve dinner from six to seven.
TONY But she said they were serving until eight o'clock. And she also said breakfast started at seven-thirty...
CHRIS ... when, in fact, it says here that breakfast starts at seven.
TONY And she said that Kingscombe was the next hostel, but it isn't. It's Charlestown.
CHRIS And finally she said that Kingscombe was ten kilometres away. But it is fifteen kilometres.

3 Aim: to check activity 2; to practise listening for specific information.

- Play the tape and ask the students to check their answers to 2.

VOCABULARY AND WRITING

1 Aim: to present the words in the vocabulary box.

- Ask the students to check they know what the words mean. Ask them to look for the words in the sentences they first saw them.

2 Aim: to practise writing.

- Ask the students to complete the letter using reported speech and as many of the words in the vocabulary box as possible.

- You may like to ask the students to do this activity for homework.

4 Listen and number the next part of the conversation in the order you hear it.

TONY And she said the last bus left at nine o'clock. But it leaves at eight o'clock. ☐
CHRIS But, in fact, it's three kilometres. ☐
TONY Yes, and she said it was two kilometres to the city centre. ☐
CHRIS It's very strange. She said one night's stay cost £6.50, but it costs £6.15. ☐

GRAMMAR

Reported speech: statements

You report what people said by using *said (that)* + clause. If the tense of the verb in the direct statement is the present simple, the tense of the verb in the reported statement is the past simple. Pronouns also change.

Direct statement	Reported statement
*'The last bus **leaves** at nine o'clock,' he said.*	***He said that** the last bus **left** at nine o'clock.*
*'**It's** two kilometres to the city centre,' he said.*	***He said** it **was** two kilometres to the city centre.*

Other tenses

Other tenses 'move back' in reported speech.

*'I've just **started** work at the hostel,' he said.*
***He said** he **had** just **started** work at the hostel.*
*'We**'re going** to Oxton,' he said.*
***He said** they **were going** to Oxton.*
*'We**'ll stay** for just one night,' he said*
***He said** they **would stay** for just one night.*
*'I **started** work last Monday,' he said.*
***He said** he **had started** work last Monday.*

1 Write what the people actually said.

1 Tony said they were both sixteen.
2 The receptionist said it took an hour on foot.
3 She said there was a bus every hour.
4 She said there wasn't much to do in the evening.
5 Chris said they were very tired.
6 He said they needed an early night.

1 'We're both sixteen', said Tony.

2 Complete the rest of the conversation in *Listening and reading* activity 4.

CHRIS And she said the bus _____ fifteen minutes. But in fact, it takes _____ minutes.
TONY And she said the hostel _____ at eleven am, but it _____ open all day.
CHRIS It seems that they _____ dinner from six to seven.
TONY But she said they were serving until eight o'clock. And she also said breakfast _____ at seven-thirty....
CHRIS ... when, in fact, it says here that breakfast _____ at seven.
TONY And she said that Kingscombe _____ the next hostel, but it _____. It's Charlestown.
CHRIS And finally she said that Kingscombe _____ ten kilometres away. But it _____ fifteen kilometres.

3 Listen and check.

VOCABULARY AND WRITING

1 Here are some new words from this lesson. Check you know what they mean.

youth hostel camping no smoking car park facilities charges adult travel

2 Complete the letter Chris and Tony wrote to the manager of Canbury Youth Hostel. Use as many of the words in the vocabulary box as possible.

9, King Street
Shrewsbury
SY2 6HJ
16 October, 1996

The Manager
Canbury Youth Hostel
Jackson Lane
Canbury

Dear Sir,
I'm writing to complain about the information your receptionist gave us when we stayed at the youth hostel. First of all, she said that one night's stay cost £6.50 when in fact, it costs £6.15. Then, she said that it was two ...

40 Dear Jan ... Love Ruth

Tense review

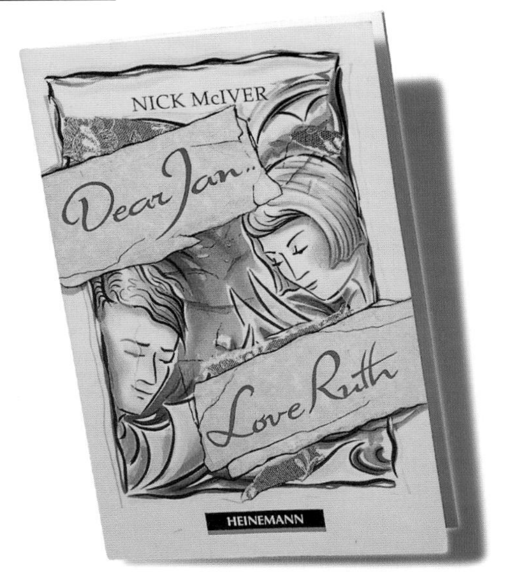

GRAMMAR

> Tense review
> **There are five tenses presented in *Reward* Elementary.**
> **Present simple**
> *I'**m** Polish. My name's Jan. What'**s** your name?*
> **Present continuous**
> *You'**re staying** with Mr and Mrs Hawkins. Mario **is** also **living** with them.*
> **Past simple**
> *A young man **arrived** at Brighton station.*
> *His name **was** Jan Polanski.*
> **Present perfect**
> *I'**ve told** him to go away.*
> *My parents **haven't met** many foreigners.*
> **Future simple**
> *I'**ll see** you tomorrow.*
> *I'**ll miss** you.*

1 Match the tenses 1 – 5 with their forms a – e below.

1 present simple
2 present continuous
3 past simple
4 present perfect
5 future simple

a most regular verbs: infinitive + *-ed*
b infinitive, or infinitive + *-s* for third person singular
c *am/is/are* + present participle
d *has/have* + past participle
e *will* + infinitive

2 Write the name of the tenses in 1 by their uses below.

a talking about recent events, such as a past action which has a result in the present
b saying what is happening now or around now
c making predictions
d talking about present customs and routines
e talking about finished actions in the past
f talking about experiences
g making decisions at the time of speaking

READING AND LISTENING

1 You're going to read a story called *Dear Jan ... Love Ruth*, by Nick McIver. Look at the cover of the book. What type of story do you think it is?

– a love story – a detective story – science fiction
– a mystery

2 Work in pairs. Part 1 of the story is called *The arrival*. Here are some words from part 1. Can you predict what happens?

Jan Polanski Poland language school Brighton stay family disco girls boyfriends stepped foot pretty name Ruth dance

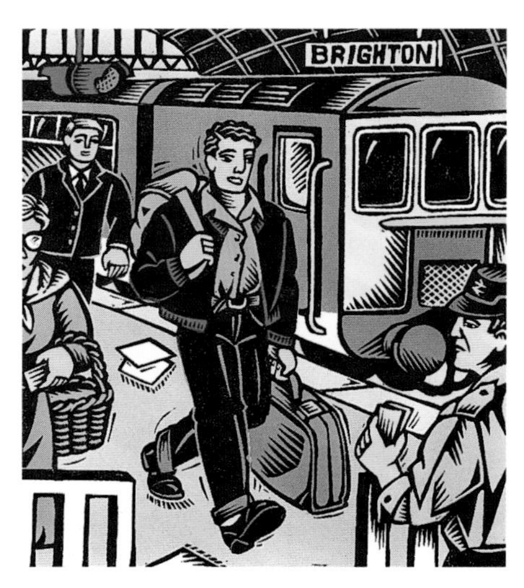

40

GENERAL COMMENTS

Tense review
This final teaching lesson uses a story to review all the tenses presented in *Reward* Elementary.

Stories and readers
Most students enjoy stories and if they are happy to read or listen to them, they should usually be left to enjoy them. The activities set in this story are designed not so much to test comprehension as to help the students understand the story better. It's also important to break up longer stories so the students are not daunted by long blocks of text. This story is in five parts, each with different activities.

Dear Ruth ... Love Jan comes from the Elementary level of the *Heinemann Guided Readers.* In this lesson it is in a shortened form, and if your students enjoy the story they may well enjoy the longer version. There are other stories at a suitable level in this series, and you may want to encourage your students to do some extensive reading using this source of material.

GRAMMAR

1 Aim: to review the tenses and their forms.

- Ask the students to read the information in the grammar box and then to do the exercises.
- There should be no surprises for the students in this activity, so you may like to do it quickly with the whole class.

Answers
1b 2c 3a 4d 5e

2 Aim: to focus on the tenses and their uses.

- Explain that one tense may have more than one use.
- Ask the students to do this activity on their own and then to check it with another student.
- Check the answers with the whole class.

Answers
a present perfect
b present continuous
c future simple
d present simple
e past simple
f present perfect
g future simple

READING AND LISTENING

1 Aim: to focus on text type; to prepare for reading.

- Explain that there may be a lot of information to be gained from looking at the illustrations and establishing the text type. This may eventually help them to get a better picture of the main idea of the story.

Answer
a love story

2 Aim: to prepare for reading; to pre-teach difficult vocabulary.

- In fact, this vocabulary may not be too difficult, but as it is contained in the first part of the story and as its meaning needs to be clear for the students to understand this first part, then it is better to provide an opportunity to pre-teach it. Ask students to explain any words which others find difficult before you explain it yourself.
- Ask the students to work in pairs and to predict what the first part of the story is about.
- Ask several pairs to give feedback to the rest of the class.

3 Aim: to practise reading for main ideas.

- Ask the students to read the first part of the story and to check their answers to 2.
- Did everyone predict correctly?

4 Aim: to prepare for reading.

- Once again, ask the students to predict what the next part of the story is likely to be about.

5 Aim: to practise reading for main ideas.

- Ask the students to read the next part of the story. Did they predict correctly in 4?

Answers
1 yes
2 by the sea
3 her ex-boyfriend
4 go outside
5 yes

6 Aim: to prepare for listening.

- Ask the students to say what they think Ruth's parents will be like. Do they think her parents will approve of Jan?
- Ask the students to decide who they think is speaking. It will be difficult to distinguish at this stage between Mr and Mrs Clark, but the recording will make it clear.
- Play the tape and ask them to check.

Answers
1 Jan 2 Jan 3 Mr Clark 4 Jan 5 Ruth
6 Mrs Clark 7 Mrs Clark 8 Ruth 9 Ruth

7 Aim: to prepare for listening.

- Ask the students to say what will happen next.
- At this stage, it is to be hoped that the students will be enjoying the story. As a result, the preparation work doesn't need to take so much time as in earlier stages of this lesson, and the tasks set while listening and reading do not need to be so elaborate or supportive.

8 Aim: to practise reading for main ideas.

- Ask the students to turn to the communication activity for the next part of the story.

9 Aim: to predict the ending of the story.

- Ask the students to say if the story has a happy or a sad ending. Ask them also if they have enjoyed the story so far.
- Ask the students to turn to the communication activity as instructed, if they would like to know the ending of the story.

VOCABULARY AND WRITING

1 Aim: to check comprehension of the words in the vocabulary box.

- Ask the students to check they understand the words and to try and remember where they first saw them.

2 Aim: to practise writing.

- Ask the students to write a different ending to the story. There are many places where their ending could begin. It could begin, for example, after Jan meets Ruth's parents.
- You may like to ask the students to do this activity for homework.
- Ask the students to read out their endings to the rest of the class.

3 Read part 1 and find out if you guessed correctly in 2.

The arrival

A young man arrived at Brighton station. His name was Jan Polanski and he came from Poland. He was in England for a course at an English language school. He took a taxi to the Modern Language Institute, went inside and met the director.

'Welcome to Brighton,' the director said. 'You're staying with the Hawkins family. Ah! Here's Mario. He's also living with them.'

'Hello, Jan,' said Mario.

That evening, after dinner Mario said, 'Would you like to come to a disco next Saturday?'

'Yes,' said Jan. 'Thanks very much.'

On Saturday Jan went to Mario's room. He was ill. 'I can't go to the disco tonight, Jan,' said Mario. 'But here's the address.'

Jan arrived at the disco at nine o'clock. He liked dancing, but most of the girls were with their boyfriends. Suddenly a girl stepped on his foot.

'Oh,' she said. 'I'm sorry.'

'That's all right,' said Jan. He looked at the girl. She was very pretty. 'Can I buy you a drink?' asked Jan. They went to the bar.

'You're not English, are you?' said the girl.

'No,' said Jan. 'I'm Polish. My name's Jan. What's your name?'

'Ruth,' she said. 'Ruth Clark.'

'Would you like to dance?' said Jan.

'Yes,' said Ruth.

4 What do you think happens next? Work in pairs and guess the answers to these questions about part 2.

1 Will they see each other again?
2 Where do they go?
3 Who does Ruth see?
4 What does she tell him to do?
5 Do Ruth and Jan like each other?

5 Read part 2 and find the answers to the questions in 4.

Jan and Ruth

The next day, Jan met Ruth and they walked by the sea. Then they went to a coffee bar. Suddenly a tall man came over to the table.

'Jan, can you go outside?' said Ruth.

Jan waited outside for about ten minutes. Then the man came out and walked away.

'Who was that?' asked Jan.

'That was Bill. He was my boyfriend. I've told him to go away. I don't like him any more.' She looked into Jan's eyes. 'Jan, I ... like you ... very much.'

Jan smiled. 'I like you very much too,' he said.

6 Work in pairs. Part 3 is called *Ruth's parents*. Here are some sentences from part 3 of the story. Decide who is speaking and to who.

1 'How do you do, Mr and Mrs Clark, ...'
2 'Sugar, but no milk.'
3 'These foreigners have strange ideas, ...'
4 'Your parents don't like me very much.'
5 'My parents haven't met many foreigners.'
6 'Well, he didn't speak English very well.'
7 'What's wrong with an English boyfriend?'
8 'But I don't like Bill any more, ...'
9 '... but I do like Jan. Maybe I love him.'

Now listen and check.

7 Work in pairs. Part 4 is called *Going home*. Here are some phrases from part 4. What do you think happens next?

last day goodbye sad come to Poland living room Bill was there next morning station I'll miss you I love you train started to leave

8 Turn to Communication activity 18 on page 103 and read part 4.

9 Work in pairs. Do you think the story has a happy or a sad ending? Talk about what will happen to Jan, Ruth ... and Bill.

If you'd like to know the ending, turn to Communication activity 24 on page 104.

VOCABULARY AND WRITING

1 Here are some words from the story. Check you remember what they mean.

language school director boyfriend step sea coffee bar foreigner strange miss forget love

2 Write a paragraph describing a different ending to the story.

In December, Ruth bought a train ticket to Poland...

Progress check 36–40

VOCABULARY

1 You put a preposition after many verbs.

listen to   agree with   decide to   laugh at

Match the verbs and the prepositions. Use your dictionary if necessary.

Verbs – apologise   belong   complain   go   hear   insist   pay   talk   think   worry
Prepositions – for   to   about   with   of   for   on   in

2 Here are some of the topics in *Reward* Elementary.

countries   jobs   family   furniture   entertainment   means of transport   shops   food and drink   clothes   sports

Try to think of two words which go with each topic.

3 Work in pairs and compare your answers to 2.

GRAMMAR

1 Make decisions about the following situations.

1 you've got nothing to eat
2 you're very tired
3 you don't feel well
4 you can't remember what a word means
5 you haven't got any money
6 you haven't spoken to your friend for a few days

1 I'll go shopping.

2 Make predictions about the following things.

1 tomorrow's weather
2 traffic in your town
3 the next World Cup
4 the next government of your country
5 your life in ten years' time
6 your English lessons

3 Rewrite these sentences in the passive.

1 Michelangelo painted the Sistine Chapel.
2 They grow cotton in Egypt.
3 They make Mercedes cars in Germany.
4 Hemingway wrote *The Old Man and the Sea*.
5 Verdi composed *Aida*.
6 William I built the Tower of London.

4 Rewrite these sentences in reported speech.

1 'I'm ill,' he said.
2 'It closes at seven,' she said.
3 'It leaves in five minutes,' she said.
4 'I work in an office,' he said.
5 'We live in London,' they said.
6 'She goes shopping on Saturday,' he said.

5 Write a sentence about yourself using each of these tenses.

1 present simple
2 present continuous
3 past simple
4 present perfect
5 future simple

SOUNDS

1 Say these words aloud. Is the underlined sound /ɔː/ or /ɔɪ/?

toy   tore   boy   bore   noise   raw   door   war   more

Listen and check. As you listen, say the words aloud.

Progress check 36 – 40

GENERAL COMMENTS

You can work through this Progress check in the order shown, or concentrate on areas which have caused difficulty in Lessons 36 - 40. You can also let the students choose the activities they would like or feel the need to do.

VOCABULARY

1 Aim: to focus on prepositions after verbs.

- The students will have come across prepositions of time and place. They will also have noticed some multi-part verbs, i.e. verbs which have more than one word. Without going into the complex area of phrasal verbs, this activity focuses on some of the important verbs which are followed by a preposition.
- Ask the students to match the verbs and the prepositions. Explain that some verbs may go with more than one preposition.

Answers
apologise for/to, belong to, complain about/to, go to/in, hear of/about, insist on, pay for, talk about/to, think about, worry about

2 Aim: to revise vocabulary and topics.

- Remind the students that recording new words under the topics they belong to will be helpful when they need to revise them.
- Ask the students to revise the vocabulary they have come across in *Reward* Elementary by asking them to think of or to find words which go under the topic headings suggested. They may find the Map of the book helpful.

3 Aim: to practise talking; to revise vocabulary.

- Ask the students to check their answers to 2 in pairs.

GRAMMAR

1 Aim: to revise the future simple for decisions.

- Ask the students to write sentences saying what they'll do in the situations mentioned. Remind them to use the future simple.

Possible answers
1 I'll go shopping.
2 I'll go to bed.
3 I'll go to the doctor.
4 I'll look it up in the dictionary.
5 I'll ask a friend to give me some.
6 I'll call him.

2 Aim: to revise the future simple for predictions.

- Ask the students to make their own predictions about the things mentioned.

3 Aim: to revise the passive.

Answers
1 The Sistine Chapel was painted by Michelangelo.
2 Cotton is grown in Egypt.
3 Mercedes cars are made in Germany.
4 *The Old Man and the Sea* was written by Hemingway.
5 *Aida* was composed by Verdi.
6 The Tower of London was built by William I.

4 Aim: to revise reported speech.

Answers.
1 He said he was ill.
2 She said it closed at seven.
3 She said it left in five minutes.
4 He said he worked in an office.
5 They said they lived in London.
6 He said she went shopping on Saturday.

5 Aim: to revise the tenses presented in *Reward* Elementary.

- Ask the students to write true sentences about themselves using all the tenses.

SOUNDS

1 Aim: to focus on /ɔɪ/ and /ɔː/.

- Write the two phonemes on the board.
- Say the words aloud and ask students to come and write the words on the board.

Answers
/ɔɪ/: toy, boy, noise
/ɔː/: tore, bore, raw, door, war, more

- Play the tape and ask the students to listen and check.
- Point out the different spelling of the two sounds.

2 Aim: to focus on stressed syllables in words.

- Write the words on the board.
- Ask students to come to the board and underline the stressed syllable.

Answers
invent, compose, discover, adult, travel, director, foreigner, forget

- Play the tape and ask the students to repeat the words.

3 Aim: to focus on stressed words in sentences.

- Remind the students that the speaker will stress the words that he or she considers to be important.
- Play the tape and ask the students to underline the stressed words. You may need to play the tape several times.

Answer
A young man arrived at Brighton station. His name was Jan Polanski. He came from Poland. He was in England for a course at an English language school. He took a taxi to the Modern Language Institute. He went inside and met the director.

- Ask the students to read the passage aloud. Make sure they stress the correct words.

SPEAKING

Aim: to revise all the language presented in *Reward* Elementary.

- This is designed to be a light-hearted end to the whole course. You may want to make enlarged photocopies of this page.
- Make sure you bring a few counters and some dice to class.
- You may like to encourage the groups to race against each other. So the first person to finish wins in the group, and the first group to finish wins in the class.

2 Underline the stressed syllable in these words.

invent compose discover adult travel director foreigner forget

Listen and check. As you listen, say the words aloud.

3 Listen and underline the words the speaker stresses.

A young man arrived at Brighton station. His name was Jan Polanski. He came from Poland. He was in England for a course at an English language school. He took a taxi to the Modern Language Institute. He went inside and met the director.

Now read the passage aloud. Make sure you stress the same words.

SPEAKING

You're going to revise the grammar and vocabulary you have learnt in *Reward* Elementary by playing *Reward Snakes and Ladders*. Work in groups of three or four and follow the instructions.

Reward Snakes and Ladders

1. Look at the game board on Communication activity 1 on page 98.
2. Each player puts their counter on the square marked START and throws the dice.
3. The first player to throw a six starts.
4. Each player then throws the dice and moves his/her counter along the board according to the number thrown on the dice. As each player lands on a square, he/she has to answer the question on the square. (You can look back at the lesson to help you.) If a player answers the question correctly, he/she can remain on the square until their next turn. If a player answers a question incorrectly, he/she must go back 3 squares. The other players decide whether the answer given is right or wrong. If you land on a Progress Check square, the player on your left can ask you any question they like.
5. If you land on a ladder, you go up to the square shown and answer the question. If you land on a snake, you go down to the square shown and answer the question.
6. The winner is the first person to reach FINISH.

Communication activities

1 *Progress check 36–40*

Speaking

Now play *Reward Snakes and Ladders.*

FINISH

Progress check 36 – 40 ?

40 Where's Jan staying? Have Ruth's parents met many foreigners?

39 When are the meals at the youth hostel?

38 Where is coffee grown? What do they grow in your country?

37 What will the weather be like tomorrow?

36 You need to get to London quickly. What will you do?

31 What must you or mustn't you do in your English lesson?

32 Have you ever been to London? Have you ever stayed in a hotel?

33 What's happened to Barry? What's happened to you today?

34 What's your advice for a perfect day out?

35 How did Beate pass her exam? How does Antonia play the guitar?

Progress check 31 – 35 ?

Progress check 26 – 30 ?

30 What's your favourite sport?

29 Is Thailand smaller than Britain? Is your country colder than Britain?

28 Name three parts of the body.

27 What's Joan's bag made of? When did she lose it?

26 Do you usually go shopping by yourself?

21 Why did Agatha Christie disappear?

22 When is New Year's Day? When is your birthday?

23 What's Harriet doing? What are you wearing?

24 What's Fiona going to do in the New Year? What's your New Year's resolution?

25 What would you like to eat? Do you like pasta?

Progress check 21 – 25 ?

Progress check 16 – 20 ?

20 Did Mary and Bill stay with friends? Where did you go on holiday?

19 What's Nick like? What are you like?

18 Where was Sting born? Where did Whitney Houston live as a child?

17 Do Jean and Tony need any fruit and vegetables? What do you need to buy?

16 Who was your first teacher? Who was your first friend?

11 What does Tanya Philips do in the evening? What do you usually do in the evenings?

12 How does Katie Francis get to work? How do you get to school/work?

13 Can cats swim? Can you drive?

14 Where's the bank? How do you get to the station?

15 Where are Janet and the kids staying? What are you doing at the moment?

Progress check 11 – 15 ?

Progress check 6 – 10 ?

10 Does Octavio like dancing? Do you like jazz?

9 How does Otto relax? How do you relax?

8 Where's the kitchen in your home? Have the Kapralovs got any chairs?

7 What's the time? What's the time in Britain?

6 What's Jenny's father's name? What's your mother's name?

1 Where is Marie from? Where are you from?

2 What's Michiko's job? What's your job?

3 Are you married? Is Greg Sheppard a doctor?

4 How many students are there in your school?

5 Where's your bag? Have you got a watch?

Progress check 1 – 5 ?

START

2 *Lesson 5*

Speaking and listening, activity 4

Look at this picture for 30 seconds.

Now turn back to page 11.

4 *Lesson 14*

Grammar and functions, activity 5

Student A: Look at the map below. Tell Student B where each shop and town facility is on your map.

Now listen to Student B and mark each shop and town facility he/she describes.

When you've finished, show your map to your partner. Have you marked the map correctly?

3 *Lesson 18*

Listening and speaking, activity 1

Student A: Listen and find out:

- where Whitney Houston was born
- who she started singing with
- how many copies of her first two albums she sold
- when she appeared in *The Bodyguard*

Now turn back to page 43.

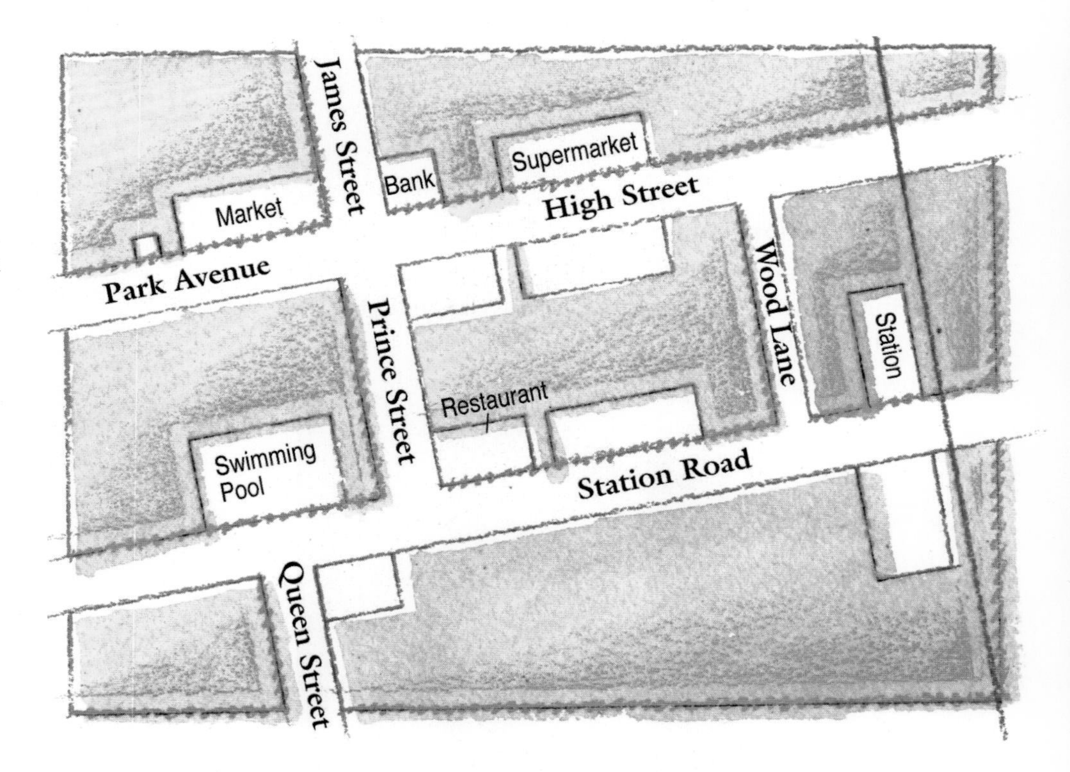

Now turn back to page 33.

5 *Lesson 22*

Speaking, activity 2

Pair A: Rewrite these facts as quiz questions.

1 The Kobe earthquake was in 1995.
2 Martin Luther King died in 1968.
3 Alexander Graham Bell invented the telephone.
4 Michelangelo was 88 years old when he died.
5 Joseph Stalin was born in Georgia.

1 When was the Kobe earthquake?

Now continue the quiz with Pair B. Ask and answer each other's questions.

6 *Lesson 21*

Grammar, activity 3

Write five statements about your past, three true and two false.

I was born in Belgium. I married Hercule Poirot.

Now work in pairs. Show your statements to your partner. Your partner must try to guess which are the false statements.

You didn't marry Hercule Poirot!

Now turn back to page 51.

7 *Lesson 23*

Functions and grammar, activity 4

Student A: Describe the photo to Student B. Now listen to Student B's description of a photo. Is the photo the same or different?

Now turn back to page 55.

8 *Lesson 21*

Vocabulary and reading, activity 5

Student A: Ask Student B these questions. Answer his/her questions in turn.

1 Why was Agatha Christie famous?
2 What was the final mystery?
3 When was she born?
4 Where did she live?
5 Who did she marry in 1914?

Now turn back to page 51.

9 *Lesson 36*

Reading and speaking, activity 3

Student B: You want to know more about the trip from the Cape to Victoria Falls. Ask Student A, who is the travel agent, the following questions:

Will there be a guided tour in Cape Town?
Will you have time to relax at the Victoria Falls?
Will the beds on the train be double or single beds?

With Student C, decide what you'll do.

10 *Lesson 2*

Vocabulary and sounds, activity 2

Look at the pictures for different jobs and check you know what the jobs are.

student

receptionist

secretary

teacher

farmer

Now turn back to page 4.

11 *Lesson 36*

Reading and speaking, activity 3

Student C: You want to know more about the trip from the Cape to Victoria Falls. Ask Student A, who is the travel agent, the following questions:

Will the flight to and from Cape Town be first class or tourist class?
Where will you stay in Johannesburg?
Will there be cabins with private bathrooms on the train?

With Student B, decide what you'll do.

12 *Lesson 14*

Grammar and functions, activity 5

Student B: Listen to Student A and mark the shops and town facilities he/she describes. Now tell Student A where each shop and town facility is on your map.

When you've finished, show your map to your partner. Have you marked the map correctly?

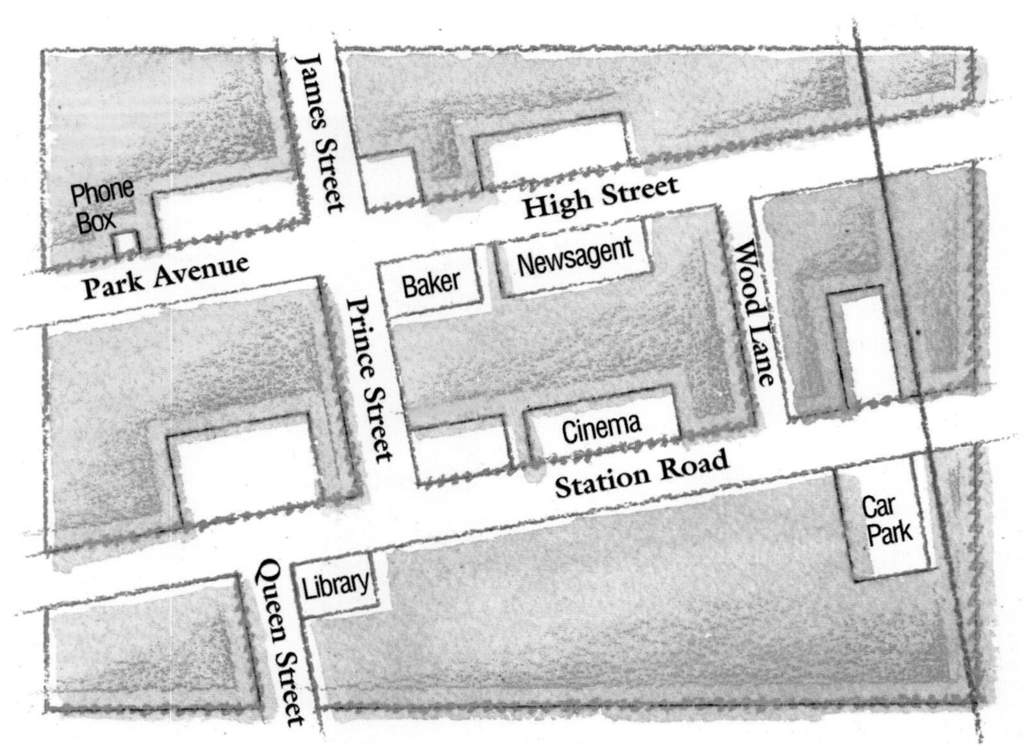

Now turn back to page 33.

13 *Lesson 23*

Vocabulary and listening, activity 1

Look at the photo for different items of clothing and check you know what the words mean.

Now turn back to page 54.

14 *Lesson 6*

Vocabulary and sounds, activity 2

Look at the photo of Holly's family. Did you guess correctly in activity 1?

1. Andrew, brother
2. Antonia, Andrew's wife
3. David, grandfather
4. Philip, father
5. Steve, brother
6. Kate, grandmother
7. Jenny, mother
8. Holly, daughter

Now turn back to page 14.

15 *Lesson 21*

Vocabulary and reading, activity 5

Student B: Ask Student A these questions. Answer his/her questions in turn.

1. When did she write her first story?
2. What did she do in December 1926?
3. What did everyone think?
4. Where did her husband find her?
5. Who did she marry in 1930?

Now turn back to page 51.

16 *Lesson 28*

Functions and grammar, activity 3

Student A: Listen to Student B and say what he/she *should/shouldn't* do.

Now act out these situations with Student B. Listen to his/her advice.

– you feel sick
– you feel unhappy all the time
– you are hungry
– you don't like your job

Now turn back to page 66.

17 *Lesson 12*

Vocabulary and reading, activity 1

Look at the photos of different means of transport and check you know what the words mean.

train

bicycles

ferry

plane

underground

tram

car

18 *Lesson 40*

Reading and listening, activity 8. Part 4

Going home

It was the last day of the course at the Modern Language Institute and Jan was very sad. He said goodbye to Mario and his other friends and left the school. That night, Jan and Ruth went for a long walk in Brighton.

'I love you, Ruth,' Jan said.

'I love you too, Jan.'

'I'm going home tomorrow. But why don't you come to Poland at Christmas?' said Jan.

'Yes,' said Ruth. 'I'd love to.'

Jan suddenly laughed. 'I'm going to see you again!'

Ruth got home at eleven o'clock that evening. She went into the house and her mother met her in the hall.

'You've got a visitor, Ruth,' she said.

Ruth went into the living room. Bill was there.

The next morning, Ruth went to Brighton station with Jan.

Jan said, 'It's September now. And you're coming to Poland in December.'

'I know,' said Ruth. 'But I'll miss you.'

At that moment, the train started to leave.

'Goodbye, Ruth,' said Jan. 'I love you.'

'Goodbye, Jan. Write to me.'

'Yes, of course.'

Ruth walked away from the station. She went down to the sea and thought about Jan.

Now turn back to page 95.

19 *Lesson 36*

Reading and speaking, activity 3

Student A: You're a travel agent. Student B and C want to know more about the trip from the Cape to Victoria Falls. Read the information below and answer their questions.

- there will be a personal guided tour in Cape Town or a large group tour
- they will be able to stay in a hotel or on the train while in Johannesburg
- there will be time to relax at the Victoria Falls or they will be able to go to the game reserve
- the flight to and from Cape Town will be first class or tourist class
- the beds on the train will be double beds or two single beds
- there will be cabins on the train with or without bathrooms

Ask them to make a decision.

20 *Lesson 4*

Sounds and vocabulary, activity 2

Check your answers.

3, 13, 30, 4, 14, 40, 5, 15, 50, 6, 16, 60, 7, 17, 70, 8, 18, 80, 9, 19, 90 100

Now turn back to page 8.

21 *Lesson 28*

Functions and grammar, activity 3

Student B: Act out these situations with Student A and listen to his/her advice.

- you don't get enough exercise
- you feel tired all the time
- you are thirsty
- you have got a headache

Now listen to Student A and say what he/she *should/shouldn't* do.

Now turn back to page 66.

22 *Lesson 22*

Speaking, activity 2

Pair B: Rewrite these facts as quiz questions.

1 The Mexico City earthquake was in 1984.
2 Yuri Gagarin was the first man in space.
3 The first American walked on the moon in 1969.
4 King Henry VIII of England and Wales had six wives.
5 Sigmund Freud was born in Vienna.

1 When was the Mexico City earthquake?

Now continue the quiz with Pair A. Ask and answer each other's questions.

23 *Lesson 18*

Listening and speaking, activity 1

Student B: Listen and find out:

- when Whitney Houston was born
- when she started singing
- when she had her first hit
- who she appeared with in *The Bodyguard*
- when she had a hit with *I will always love you*

Now turn back to page 43.

24 *Lesson 40*

Reading and listening, activity 9. Part 5

The end

Jan wrote several letters to Ruth. But every time Mrs Clark found the letters she burnt them. Ruth was very sad. She thought, Jan doesn't love me any more. He's forgotten about me.

Bill was very kind to Ruth at this time. At the end of November, Ruth went to a party with Bill.

Now turn back to page 95.

25 *Lesson 23*

Reading and speaking, activity 3

Add up your scores using the following table. Then look at the profiles below to find out what your clothes say about you.

1	a 2	b 3	c 1
2	a 2	b 3	c 1
3	a 2	b 1	c 3
4	a 2	b 1	c 3
5	a 3	b 2	c 1
6	a 3	b 2	c 1
7	a 3	b 2	c 1
8	a 2	b 1	c 3
9	a 1	b 3	c 2
10	a 1	b 2	c 3

21 – 30 points. You like to wear exactly what you want. Sometimes this may get you into trouble.

11 – 20 points. You are quite casual. Sometimes you don't wear the right clothes for the situation.

1 – 10 points. You're very careful to wear the right clothes for the right situation.

26 *Lesson 13*

Vocabulary and reading, activity 4

11 – 15 points. Excellent! You're an all-rounder!

6 – 10 points. Good. You can do many different things.

1– 5 points. You're not an all-rounder! You like to specialise!

Now turn back to page 30.

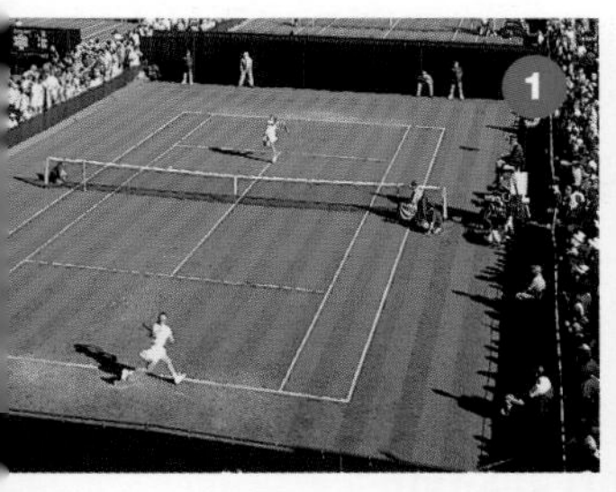

27 *Lesson 23*

Functions and grammar, activity 4

Student B: Listen to Student A's description of a photo. Now describe your photo to Student A. Is the photo the same or different?

Now turn back to page 55.

28 *Lesson 30*

Vocabulary and listening, activity 1

Look at the photos of different sports and check you know what they are.

Grammar review

CONTENTS

Present simple

Form

You use the contracted form in spoken and informal written English.

Be

Affirmative	Negative
I'm (I am)	I'm not (am not)
you we 're (are) they	you we aren't (are not) they
he she 's (is) it	he she isn't (is not) it

Questions	Short answers
Am I?	Yes, I am. No, I'm not.
Are you/we/they?	Yes, you/we/they are. No, you/we/they're not.
Is he/she/it?	Yes, he/she/it is. No, he/she/it isn't.

Have

Affirmative	Negative
I you have we they	I you haven't (have not) we they
he she has it	he she hasn't (has not) it

Questions	Short answers
Have I/you/we/they?	Yes, I/you/we/they have. No, I/you/we/they haven't.
Has he/she/it?	Yes, he/she/it has. No, he/she/it hasn't.

Regular verbs

Affirmative	Negative
I you work we they	I you don't (do not) work we they
he she works it	he she doesn't (does not) work it

Questions	Short answers
Do I/you/we/they work?	Yes, I/you/we/they do. No, I/you/we/they don't (do not).
Does he/she/it work?	Yes, he/she/it does. No, he/she/it doesn't (does not).

Question words with *is/are*

What's your name? *Where are your parents?*

Question words with *does/do*

Where does he live? *What do you do?*

Present simple: third person singular

You add *-s* to most verbs.
takes, gets
You add *-es* to *do, go* and verbs which end in *-ch, -ss, -sh* and *-x*.
does, goes, watches, finishes
You drop the *-y* and add *-ies* to verbs ending in *-y*.
carries, tries

Use

You use the present simple:

- to talk about customs. (See Lesson 7)
  In Spain people have dinner at ten or eleven in the evening.
  In Britain people leave work at five in the afternoon.
- to talk about habits and routines. (See Lesson 9)
  I go running every day.
  We see friends at the weekend.
- to say how often you do things. (See Lesson 11)
  I always get up at seven o'clock.
  I sometimes do the shopping in the evening.
- to describe something that is true for a long time. (See Lesson 23)
  He wears glasses.

Present continuous

Form

You form the present continuous with *be* + present participle (*-ing*). You use the contracted form in spoken and informal written English.

Affirmative	Negative
I'm (am) working	I'm not (am not) working
you we 're (are) working they	you we aren't (are not) working they
he she 's (is) working it	he she isn't (is not) working it

Questions	Short answers
Am I working?	Yes, I am. No, I'm not.
Are you/we/they working?	Yes, you/we/they are. No, you/we/they aren't.
Is he/she/it working?	Yes, he/she/it is. No, he/she/it isn't.

Question words

What are you doing? *Why are you laughing?*

Present participle (*-ing*) endings

You form the present participle of most verbs by adding *-ing*:
go – going *visit – visiting*
You drop the *-e* and add *-ing* to verbs ending in *-e*.
make – making *have – having*
You double the final consonant of verbs of one syllable ending in a vowel and a consonant and add *-ing*.
get – getting *shop – shopping*
You add *-ing* to verbs ending in a vowel and *-y* or *-w*.
draw – drawing *play – playing*
You don't usually use these verbs in the continuous form.
believe feel hate hear know like love see smell sound taste think understand want

Use

You use the present continuous:

- to describe something that is happening now or around now. (See Lessons 15 and 23)
  We're flying at 10,000 metres.
  She's wearing a yellow dress.
- to talk about future intentions or plans which are fairly certain. (See Lesson 24)
  We're going to see Mary on Saturday.
  He's coming round this evening.

Past simple

Form

You use the contracted form in spoken and informal written English.

Be

Affirmative	Negative
I he was she it	I he wasn't (was not) she it
you we were they	you we weren't (were not) they

Have

Affirmative	Negative
I you we they had he she it	I you we they didn't (did not) have he she it

Regular verbs

Affirmative	Negative
I you we they worked he she it	I you we they didn't (did not) work he she it

Questions	Short answers
Did I/you/we/they work? he/she/it	Yes, I/you/we/they did. he/she/it No, I/you/we/they didn't. he/she/it

Question words

What did you do yesterday? *Why did you leave?*

Past simple endings

You add *-ed* to most regular verbs.
walk – walked *watch – watched*

You add *-d* to verbs ending in *-e*.
close – closed *continue – continued*

You double the consonant and add *-ed* to verbs of one syllable ending in a vowel and a consonant.
stop – stopped *plan – planned*

You drop the *-y* and add *-ied* to verbs ending in *-y*.
study – studied *try – tried*

You add *-ed* to verbs ending in a vowel + *-y*.
play – played *annoy – annoyed*

Irregular verbs

There are many verbs which have an irregular past simple. For list of the irregular verbs which appear in **Reward Elementary** see page 114.

Pronunciation of past simple endings

/t/ *finished, liked, walked*
/d/ *continued, lived, stayed*
/ɪd/ *decided, started, visited*

Expressions of past time

(See Lesson 22)

yesterday morning/afternoon/evening
last Saturday/week/month/year
two weeks ago/six months ago

Use

You use the past simple:

- to talk about an action or event in the past that is finished. (See Lessons 16, 18, 20 and 22)
  What were you like as a child?
  I started learning English last year.
  Did they go to France last year?

Future simple (*will*)

Form

You form the future simple with *will* + infinitive. You use the contracted form in spoken and informal written English.

Affirmative	Negative
I you we they 'll (will) work he she it	I you we they won't (will not) work he she it

Questions	Short answers
Will I/you/we/they work? he/she/it/	Yes, I/you/we/they will. he/she/it/ No, I/you/we/they won't. he/she/it/

Question words

What will you do? *Where will you go?*

Use

You use the future simple:

- to talk about a decision you make at the moment of speaking. (See Lessons 26 and 36)
  I'll have the blue T-shirt.
  I think I'll go out tonight.
  I'll call back later.
- to make a prediction or express an opinion about the future. (See Lesson 37)
  There'll be more and more people in the world.
  I think it'll be hot tomorrow.

Present perfect simple

Form

You form the present perfect simple with *has/have* + past participle. You use the contracted form in spoken and informal written English.

Affirmative		Negative	
I		I	
you	've (have) worked	you	haven't (have not) worked
we		we	
they		they	
he		he	
she	's (has) worked	she	hasn't (has not) worked
it		it	

Questions	Short answers
Have I/you/we/they worked?	Yes, I/you/we/they have. No, I/you/we/they haven't.
Has he/she/it worked?	Yes, he/she/it has. No, he/she/it hasn't.

Past participles

All regular and some irregular verbs have past participles which are the same as their past simple form.

Regular: *move – moved, finish – finished, visit – visited*
Irregular: *leave – left, find – found, buy – bought*

Some irregular verbs have past participles which are not the same as the past simple form.

go – went – gone *be – was/were – been*
drink – drank – drunk *ring – rang – rung*

For a list of the past participles of the irregular verbs which appear in **Reward Elementary**, see page 114.

Been and *gone*

He's been to America. = *He's been there and he's back here now.*
He's gone to America. = *He's still there.*

Use

You use the present perfect simple:

- to talk about past experiences. You often use it with *ever* and *never*. (See Lesson 32)
  Have you ever stayed in a hotel? (=Do you have the experience of staying in a hotel?)
  Yes, I have. (=Yes, I have stayed in a hotel at some point, but it's not important when.)
  No, I've never stayed in a hotel.

  Remember that if you ask for and give more information about these experiences, actions or states, such as *when, how, why* and *how long,* you use the past simple.
  When did you stay in a hotel? *When I was in France last year.*
- to talk about recent events, such as a past action which has a result in the present. You often use it to describe a change. (See Lesson 33)
  He's hurt his back.
  He's lost his wallet.

  You use *just* if the action is very recent.
  He's just lost his wallet.

  You use *yet* in questions and negatives to talk about an action which is expected. You usually put it at the end of the sentence.
  Have you been to the doctor yet?

  Remember to use the past simple to say when the action happened.
  He hurt his back this morning.

Questions

You can form questions in two ways:

- without a question word. (See Lesson 3)
  Are you British?
  Was he born in Italy?
  Do you have any brothers?
  Did you get up late this morning?
- with a question word such as *who, what, where, when, how* and *why*. (See Lesson 9)
  What's his job?
  How old is he?
  What do you do to relax?
  Where were you born?

You can put a noun after *what* and *which*.
What time is it? *Which road will you take?*

You can put an adjective or an adverb after *how*.
How much is it? *How long does it take by car?*
How fast can you drive?

You can use *who, what* or *which* as pronouns to ask about the subject of the sentence. You don't use *do* or *did*.
What's your first name? *Who was Agatha Christie?*

You can use *who, what* or *which* as pronouns to ask about the object of the sentence. You use *do* or *did*.
What did Agatha Christie do? *Who did she marry?*

You can form more indirect, polite questions with one of the following question phrases.
Can I help you?
Could I have some water, please?
Would you like a regular or a large Coke?

Imperatives

The imperative has exactly the same form as the infinitive (without *to*) and does not usually have a subject. You use the imperative:

- to give directions. (See Lesson 14)
  Go along East Street.
  Turn left into Prince Street.
- to give instructions and advice. (See Lesson 34)
  Come in.
  Sit down.
  Check the weather forecast before you go.

You use *don't* + imperative to give a negative instruction.
Don't take too much to carry.

Verb patterns

There are several possible patterns after certain verbs which involve *-ing* form verbs and infinitive constructions with or without *to*.

-ing form verbs

You can put an *-ing* form verb after certain verbs.
(See Lesson 10)
I like playing football on the beach.
Peter hates travelling by plane.

Remember that *would like to do something* refers to an activity at a specific time in the future.
I'd like to go to the cinema next Saturday.

When you *like doing something,* this is something you enjoy all the time.
I like going to the cinema. I go most weekends.

to + infinitive

You can put *to* + infinitive after many verbs.
Here are some of them:
decide go have hope learn like need want
He decided to go to Spain for a holiday.

Use
You use *to* + infinitive (the infinitive of purpose):

- to say why people do things. (See Lesson 34)
  You go to a chemist to buy suncream.
  You go to the bus stop to catch a bus.
- to describe the purpose of something. (See Lesson 34)
  You use ice to keep things cold.

Have got

Form
You use the contracted form in spoken and informal written English.

Affirmative		Negative	
I		I	
you	've (have) got	you	haven't (have not) got
we		we	
they		they	
he		he	
she	's (has) got	she	hasn't (has not) got
it		it	

Questions	Short answers
Have I/you/we/they got?	Yes, I/you/we/they have. No, I/you/we/they haven't.
Has he/she/it got?	Yes, he/she/it has. No, he/she/it hasn't.

Use

You use *have got* to talk about possession.
He's got a new car.
Have got means the same as *have*. You use it in spoken and informal written English.
She's got a house in London. (= She has a house in London.)

Going to

You use *going to* + infinitive:

- to talk about future intentions or plans which are fairly certain. (See Lesson 24)
  I'm going to see my friends more often.
- to talk about something that we can see now is sure to happen in the future. (See Lesson 24)
  She's going to have a baby.

Modal verbs

The following verbs are modal verbs.
can could must should will would needn't

Form
Modal verbs:

- have the same form for all persons.
  I must go. He must be quiet.
- don't take the auxiliary *do* in questions and negatives.
  Can you use a computer?
  You mustn't be late for the meeting.
- take an infinitive without *to*.
  I can type.
  You should see a doctor.

Use
You use *can*:

- to talk about general ability, something you are able to do on most occasions. (See Lesson 13)
  I can play the piano.
  I can drive a car.
- to ask for permission. (See Lesson 26)
  Can I try this on?
- to say if you're allowed to do something (See Lesson 31)
  You can go to London on Saturday.

You can also use *could*. *Can* is a little less formal than *could*.

You use *could*:

- to ask for something politely.
  Could I have some water, please?
- to ask people to do things
  Could you tell me your name?
- to ask for permission
  Could I try this on?

You use *must*:

- to talk about something you're strongly advised to do or are obliged to do. (See Lesson 31)
  You really must stop smoking.
  It's late. I must go now.
  You must stop at a red light.

You often use it to talk about safety instructions.
You must use your seatbelt.

You use *mustn't*:

- to talk about something you're strongly advised not to do. (See Lesson 31)
  You mustn't point at people.
- to talk about something you're not allowed to do. (See Lesson 31)
  You mustn't drive through a red light.

You can also use *can't*.
  You can't drive through a red light.

Remember that you don't usually use *must* in questions.

You use *should:*

- to give less strong advice. It can also express a mild obligation or the opinion of the speaker. (See Lesson 28)
  You should go to bed.
  You shouldn't go to work.

For uses of *will* see Future simple (*will*).

You use *would like* + noun or *would like to* + infinitive:

- to offer or request something politely. (See Lesson 25)
  Would you like a drink?
  What would you like to drink?
  I'd like a burger, please.
  I'd like to go to the cinema tonight.

You can also use *like* to say what you like all the time.
I like Coke. (= always)
I'd like a Coke. (= now)

You use *needn't* to talk about something it isn't necessary to do.
You needn't do your homework tonight.

Pronouns

Subject (See Lesson 10)	**Object** (See Lesson 10)	**Possessive** (See Lesson 27)
I	me	mine
you	you	yours
he	him	his
she	her	hers
it	it	its
we	us	ours
they	them	theirs

Articles

There are many rules for the use of articles. Here are some of the most useful. (See Lessons 2 and 12)
You use the indefinite article (*a/an*):

- to talk about something for the first time.
  She works in an office in Paris.
  I get a train to work.
- with jobs.
  She's a marketing consultant.
  He's a ticket inspector.

- with certain expressions of quantity.
  I go to the cinema once or twice a month.
  There are several trains a day.

You use *an* for nouns which begin with a vowel.
an accountant, an apple

You use the definite article (*the*):

- to talk about something again.
  The office is near the Gare du Nord.
  I get the eight o'clock train.
- when there is only one.
  The Channel Tunnel.
  The Eurostar service.

Before vowels you pronounce *the* /ði:/.
You don't use any article:

- with certain expressions.
  by train by plane at work at home
- with most countries, meals, languages.
  She often goes to France.
  She lives in Britain.
  Let's have lunch.
  I speak Italian.

Plurals

You form the plural of most nouns with *-s*.
(See Lessons 4 and 6)
bag – bags, book – books, key – keys

For nouns which end in *-y*, you drop *-y* and add *-ies*.
diary – diaries, baby – babies

You add *-es* to nouns which end in *-o, -ch, -ss, -sh* and *-x*.
watch – watches, glass – glasses

There are some irregular plurals.
man – men, woman – women, child – children

Possessives

Possessive *'s*

You add *'s* to singular nouns to show possession.
(See Lesson 6)
John's mother. His teacher's book.

You add *s'* to regular plural nouns.
My parents' names are Georges and Paulette.
The boys' room.

You add *'s* to irregular plural nouns.
Their children's names are Pierre and Thierry.
The men's room.

Possessive adjectives

You can find the main uses for possessive adjectives in Lesson 2.

Form

I	you	he	she	it	we	they
my	your	his	her	its	our	their

Whose

You use *whose* to ask who something belongs to.
(See Lesson 27)
Whose bag is this?
Whose are these shoes?

Expressions of quantity

Countable and uncountable nouns

Countable nouns have both a singular and a plural form.
(See Lesson 17)
a banana – bananas, a tomato – tomatoes

Uncountable nouns do not usually have a plural form.
water, juice, wine

If you talk about different kinds of uncountable nouns, they become countable.
Beaujolais and Bordeaux are both French wines.

Some and *any*

You usually use *some* with plural and uncountable nouns in affirmative sentences when you are not interested in the exact number. (See Lessons 8 and 17)
We need some fruit and vegetables.

You usually use *any* with plural and uncountable nouns in negative sentences and questions. (See Lessons 8 and 17)
We haven't got any carrots.
Have we got any milk?

You use *some* in questions when you ask for, offer or suggest something.
How about some oranges?

Adjectives

Position of adjectives

You can put an adjective in two positions. (See Lesson 4)

- after the verb *to be*.
  The book is very interesting.
- before a noun.
  It's an interesting book.

Comparative and superlative adjectives

Form

You add *-er* to most adjectives for the comparative form, and *-est* for the superlative form. (See Lessons 29 and 30)
cold – colder – coldest *cheap – cheaper – cheapest*

You add *-r* to adjectives ending in *-e* for the comparative form, and *-st* for the superlative form.
large – larger – largest *fine – finer – finest*

You add *-ier* to adjectives ending in *-y* for the comparative form, and *-iest* for the superlative form.
happy – happier – happiest *friendly – friendlier – friendliest*

You double the final consonant and add *-er* or *-est* to adjectives of one syllable ending in a vowel and a consonant.
hot – hotter – hottest *thin – thinner – thinnest*

You use *more* for the comparative form and *most* for the superlative form of longer adjectives.
expensive – more expensive – most expensive
important – more important – most important

Some adjectives have irregular comparative and superlative forms.
good – better – best *bad – worse – worst*

With the superlative form you usually use *the* before the adjective in its superlative form.
James is the tallest person in the room.

You use a comparative + *than* when you compare two things which are different.
Thailand is bigger than Britain.

Adverbs

Formation of adverbs

You use an adverb to describe a verb. (See Lesson 35)
She speaks English fluently.
He drives carelessly.

You form an adverb by adding *-ly* to the adjective.
fluent – fluently *careless – carelessly*

If the adjective ends in *-y*, you drop the *-y* and add *-ily*.
happy – happily *easy – easily*

Some adverbs have the same form as the adjective they come from.
late, early, hard

The adverb from the adjective *good* is *well*.
She's a good writer. *She writes well.*

Position of adverbs of frequency

You usually put adverbs of frequency before the verb. (See Lesson 11)
I always get up at seven o'clock.
I often have a drink with friends.

But you put them after the verb *to be*.
I'm never late for work.

Prepositions of time and place

in, at, on, to

You use *in*:

- with seasons and months of the year.
  in winter, in September, in March
- with places. (See Lesson 5)
  in the classroom, in the photograph, in Greece
- with times of the day. (See Lesson 7)
  in the morning, in the afternoon

You use *at*:

- with certain expressions. (See Lesson 11)
  at school, at home, at work
- with times of the day. (See Lesson 11)
  at night, at seven o'clock, at the weekend

You use *on*:

- with days and dates. (See Lesson 11)
  on Sunday, on Monday morning, on 15th June

You use *to*:

- with places.
  Helena goes to London every month.

You use *from ... to*:

- to express how long something lasts. (See Lesson 11)
  The shop is open from seven to nine o'clock.

Present simple passive

You form the present simple passive with *am/is/are* + past participle. (See Lesson 38)
Fiat cars are made in Italy.

Use

You use the passive to focus on the object of the sentence.
This palace was built by Palladio in 1635.

Reported speech

Statements

You report what people said by using *said that* + clause. Notice how the tense of the verb in the direct statement moves one tense back in the reported statement.

Direct statement	Reported statement
'The film finishes at ten o'clock,' she said.	*She said the film finished at ten o'clock.*
'We're going on holiday next year,' she said	*She said they were going on holiday next year.*
'We'll catch the train to Paris,' he said.	*He said they would catch the train to Paris.*
'I watched the television all evening,' he said.	*He said he had watched the television all evening.*

Irregular Verbs

Verbs with the same infinitive, past simple and past participle

cost	cost	cost
cut	cut	cut
hit	hit	hit
let	let	let
put	put	put
read /riːd/	read /red/	read /red/
set	set	set
shut	shut	shut

Verbs with the same past simple and past participle but a different infinitive

bring	brought	brought
build	built	built
burn	burnt/burned	burnt/burned
buy	bought	bought
catch	caught	caught
feel	felt	felt
find	found	found
get	got	got
have	had	had
hear	heard	heard
hold	held	held
keep	kept	kept
learn	learnt/learned	learnt/learned
leave	left	left
lend	lent	lent
light	lit/lighted	lit/lighted
lose	lost	lost
make	made	made
mean	meant	meant
meet	met	met
pay	paid	paid
say	said	said
sell	sold	sold
send	sent	sent
sit	sat	sat
sleep	slept	slept
smell	smelt/smelled	smelt/smelled
spell	spelt/spelled	spelt/spelled
spend	spent	spent
stand	stood	stood
teach	taught	taught
understand	understood	understood
win	won	won

Verbs with same infinitive and past participle but a different past simple

become	became	become
come	came	come
run	ran	run

Verbs with a different infinitive, past simple and past participle

be	was/were	been
begin	began	begun
break	broke	broken
choose	chose	chosen
do	did	done
draw	drew	drawn
drink	drank	drunk
drive	drove	driven
eat	ate	eaten
fall	fell	fallen
fly	flew	flown
forget	forgot	forgotten
give	gave	given
go	went	gone
grow	grew	grown
know	knew	known
lie	lay	lain
ring	rang	rung
rise	rose	risen
see	saw	seen
show	showed	shown
sing	sang	sung
speak	spoke	spoken
swim	swam	swum
take	took	taken
throw	threw	thrown
wake	woke	woken
wear	wore	worn
write	wrote	written

Pronunciation guide

/ɑː/	park
/æ/	hat
/aɪ/	my
/aʊ/	how
/e/	ten
/eɪ/	bay
/eə/	there
/ɪ/	sit
/iː/	me
/ɪə/	beer
/ɒ/	what
/əʊ/	no
/ɔː/	more
/ɔɪ/	toy
/ʊ/	took
/uː/	soon
/ʊə/	tour
/ɜː/	sir
/ʌ/	sun
/ə/	better

/b/	buy
/d/	day
/f/	free
/g/	give
/h/	house
/j/	you
/k/	cat
/l/	look
/m/	mean
/n/	nice
/p/	paper
/r/	rain
/s/	sad
/t/	time
/v/	verb
/w/	wine
/z/	zoo
/ʃ/	shirt
/ʒ/	leisure
/ŋ/	sing
/tʃ/	church
/θ/	thank
/ð/	then
/dʒ/	jacket

Tapescripts

Lesson 5 Listening and vocabulary, activity 2

Conversation 1

MAN Where's my pen?
WOMAN It's on the table, near your book.
MAN Oh, I see. Thanks.

Conversation 2

MAN Have you got a mobile phone?
WOMAN Yes, I have. It's in my bag. Here you are.
MAN Thanks.

Conversation 3

WOMAN Where's my bag?
MAN What colour is it?
WOMAN It's blue.
MAN It's under your chair.
WOMAN Oh, yes. Thank you.

Lesson 5 Speaking and listening, activity 2

Q Tell me about your personal possessions, Steve. What sort of things have you got?
STEVE Well, I've got a personal stereo, I suppose that's very typical these days. And I've got a computer, like everyone else.
Q Have you got a mobile phone?
STEVE No, I haven't. I don't need one at the moment. I'm still at school.
Q And have you got a bicycle?
STEVE Yes, I have. Oh, and I've got a television and a radio in my bedroom.
Q And have you got a video too?
STEVE No, I haven't. My parents have got a video, but it's downstairs.
Q And what about a watch?
STEVE Yes, I've got a watch. Oh, I'm late for school! Bye!
Q Thanks, Steve.

Lesson 7 Reading and listening, activity 3

Q So, Tony, you're Australian, right?
TONY That's right.
Q And where do you come from in Australia?
TONY From Sydney.
Q Sydney! I've heard it's very beautiful there.
TONY I think it's *very* beautiful, but it *is* my hometown.
Q Tell me about the daily routine in Australia. What time do you get up?
TONY During the week, we get up at seven in the morning.
Q What time do children start school in the morning?
TONY It's usually about nine o'clock.
Q Nine o'clock. That's later than many countries.
TONY Yes, it is.
Q And when do they finish school?
TONY At about three in the afternoon.
Q And when do people go to work in the morning?
TONY Well, we start work at nine o'clock, so we go to work at seven-thirty or eight o'clock.
Q So they start work at nine o'clock, you say?
TONY That's right. Nine o'clock.
Q And when do you have lunch?
TONY Well, we have lunch at one o'clock. We stop work and have a sandwich usually, but our main meal is dinner, in the evening.
Q And what time do you stop work?
TONY We stop work at five in the afternoon. Actually, we leave work at five in the afternoon. We probably stop work earlier!
Q And when do you have dinner?
TONY At seven o'clock in the evening, usually. We eat outside in the garden most of the year.
Q And when do you go to bed?
TONY We go to bed at eleven or twelve at night.
Q And do you work on Saturdays and Sundays?
TONY No, we don't work at the weekend.

Lesson 8 Listening and writing, activity 1

Q So Geoff, this is your home!
GEOFF That's right. Do you like it?
Q I like it very much. It's a very nice boat. It's so quiet here on the river.
GEOFF It's noisy in the summer with the tourist boats, but in winter it's perfect.
Q How big is it?
GEOFF Well, it's a special type of boat for the canals, which are very narrow in Britain. It's called a narrow boat and it's ten metres long and about two metres wide.
Q Ten metres! Is it difficult to drive it along the canals?
GEOFF At first it's difficult, but after a while, with practice, it's quite easy.
Q But I suppose with ten metres, you have a lot of space.
GEOFF Yes, well, we're in the living room, and there's a kitchen, a bathroom and two bedrooms through there.
Q How many people live in the boat with you?
GEOFF My wife and our baby daughter, three people in all.
Q Three of you, I see. And what sort of furniture do you have?
GEOFF Well, there's a fridge and a cooker, several armchairs, a television and a shower, but there isn't room for a bath and a dishwasher. But it's quite comfortable.
Q And what's the most important item for you?
GEOFF I suppose it's my computer. Yes, it's my computer.

Lesson 9 Listening and speaking, activity 2

Q So what do you do in your free time, Helen?
HELEN Well, I like to relax with a good novel.
Q Really? You like reading then?
HELEN Yes, I do. I like newspapers and magazines but for me a good novel is the best.
Q How about you Chris?
CHRIS Well, I haven't got much time for reading. I see my friends a lot and we go to a club every Friday and Saturday.
Q And what about during the rest of the week?
CHRIS Well, I stay at home and watch television.
Q Do you watch television, Helen?
HELEN No, not very much. There's a television in the sitting room, but I only watch the news. Oh, there is one programme I like – it's called *Sueños*.
Q *Sueños?* What's that?
HELEN Well, it's a very good programme on how to learn Spanish. It's on BBC. I learn languages in my free time. Last year I learned French, and next year I'd like to do Russian.
Q That's fascinating. Thank you.

Lesson 10 Listening, activity 1

A Do you like rock music?
B No, I don't. I hate it.
A What type of music do you like? Do you like jazz?
B Yes, I do. I love it.
A Who's your favourite musician?
B Miles Davis. He's great.
A I don't like him very much. I don't like jazz.
B Oh, I like it very much. What about you? Who's your favourite singer?
A I like classical music. My favourite singer is Pavarotti. Do you like classical music?
B It's all right.

Lesson 10 Listening and writing, activity 2

JOHN Hi, I'm John, I live in Brighton in England. I'm a student at university here. I like sport, football, tennis and skiing. And my favourite music is rock music. I like the *Rolling Stones*.
KATE Hello, my name's Kate and I live in Edinburgh, in Scotland. I like winter sports like skiing and ice skating, because it gets quite cold here in Scotland. In fact, I like all sports. And I also like going to the cinema.
KEITH My name's Keith and I live in Hong Kong. I like computer games and water sports, like swimming and water polo. I go to clubs most weekends with my friends, because I like dancing.
SUSIE My name's Susie and I'm from Melbourne in Australia. I like going to the beach and dancing, and I also like swimming. I read novels and go to the theatre in my free time.

Lesson 11 Listening and writing, activity 2

Q Sam, tell me about your typical day.
SAM Well, I usually get up at about midday.
Q Midday!
SAM Yes, midday.
Q But that's...
SAM Twelve o'clock, yes, that's right. Then I...
Q But what's your job?
SAM Oh, I'm a musician. But I haven't got much work at the moment.
Q OK, so what do you do after you get up?
SAM Well, I usually have breakfast.
Q So, after breakfast...?
SAM ... after breakfast, I always meet my friends and we often play football. And then we usually have some lunch.
Q And what do you do after lunch?
SAM Well, we sometimes go to a concert or we play some music.
Q And then what? Do you go to bed?
SAM Well, I usually have dinner in a restaurant at nine or ten in the evening if I'm hungry, and then...
Q ... you go to bed?
SAM ... I always go to a club or a party.
Q And when do you go home?
SAM Four or five o'clock in the morning.
Q Four or five o'clock in the morning? You do this every day?
SAM Well, not Sundays.
Q Not Sundays. What do you do on Sunday?
SAM I always stay at home and telephone my friends.
Q Every Sunday?
SAM Yes, most people are at home on Sunday.
Q I see. And this is a typical day?
SAM Yes, well, it's typical for me.

Lesson 13 Grammar, activity 2

JO What can you do?
PAT I can run 100 metres in 15 seconds and I can use a computer.
JO So can I.
PAT But I can't cook very well.
JO Nor can I. But I can speak a foreign language.
PAT What language can you speak?
JO Spanish. Can you speak Spanish?
PAT No, I can't.

Lesson 13 Speaking and listening, activity 3

FRANK Hey Ann, look I've got that, er, I've got that list. Listen, I'll read some of them out. You can see the Great Wall of China from space.
ANN Yes, I think you can.
FRANK Yeah, so that's true.
ANN Yes.
FRANK Yes, that's true. Let's have a look at this one. Here's one. Cats can't swim.
ANN Erm... I think they can, actually.
FRANK Yes, so therefore, it's, that's false.
ANN Cats can't swim, that's, no, false.
FRANK That statement's false. Where's another... oh look here... chickens can't fly.
ANN Erm...
FRANK Can they fly?
ANN I'm not sure.
FRANK I don't think they can.
ANN No, they can't.
FRANK So, that's true?
ANN Yes.
FRANK Right, OK.
ANN Chickens can't fly.
FRANK Chickens cannot fly. Computers can write novels.
ANN I don't think they can.
FRANK No, I don't think they can.
ANN That's ridiculous.
FRANK So, what that's... er, we'll put there, we'll put no.
ANN No, false.
FRANK That's false. Cameras can't lie.
ANN Yes, they can.
FRANK Yes, so that's erm, that's false. That's false too. Erm... England can win the next World Cup competition.
ANN Oh, I don't know.
FRANK I think that's false. Do you think?
ANN I don't know. I... they could, couldn't they?
FRANK Well, they could win it.
ANN Yes. So, yes true.
FRANK Alright, we'll make it true. Right, we'll do that. Thin people can't swim very well.
ANN That's rubbish.
FRANK That's rubbish, they, er... they swim very well, so that's false. Erm... here's, here's a good one. You can never read a doctor's handwriting.
ANN Well, it's a bit of a cliché that, isn't it?
FRANK Yeah, it is.
ANN But I think it's false.
FRANK I think it's false, too. No. Erm, here we go. Oh, here's a good one. You can clean coins with Coke.
ANN I believe you can.
FRANK Yes, I've heard that too.
ANN I've never done it.
FRANK So, erm... yeah that's true.
ANN Let's say that's true.
FRANK OK, and the last one I've got here is, cats can see in the dark. I think that's erm...
ANN I think that's true.
FRANK I think it's true too. Yes, that's true.
ANN OK, let's say true.

Lesson 14 Vocabulary and listening, activity 3

Conversation 1

WOMAN Er, can you help me?
MAN Yes, of course.
WOMAN Where is the bank?
MAN There's a bank in Valley Road.
WOMAN How do I get there?
MAN Go along Prince Street, turn left up George Street. Turn left into Valley Road and the bank is opposite the cinema.
WOMAN Thank you.

Conversation 2

WOMAN Excuse me, where's the baker's?
MAN There's a baker's next to the cinema.
WOMAN How do I get there?
MAN Go up East Street and turn right into Valley Road. The baker's is on your left, opposite the florist.
WOMAN Thank you.

Conversation 3

WOMAN Can I help you?
MAN Er, yes. Where can I buy some aspirin?
WOMAN There's a chemist in Prince Street.
MAN Oh, how do I get to Prince Street?
WOMAN Go down George Street and turn right into Prince Street. The chemist is opposite the restaurant.
MAN Thank you.

Conversation 4

MAN Excuse me, where can I buy a newspaper?
WOMAN There's a newsagent in East Street.
MAN How do I get to East Street?
WOMAN Go along Prince Street. Turn right into East Street the newsagent is on your left.
MAN Thank you.

Lesson 15 Listening, activity 2

CAPTAIN Ladies and gentlemen, this is your captain speaking. I hope you're enjoying your flight to Rome this morning. At the moment, we're flying over the beautiful city of Zurich, in the centre of Switzerland. If you're sitting on the left-hand side of the plane, you can see the city from the window. We're flying at 12,000 metres and we're flying at a speed of 750 km/h. I'm afraid the weather in Rome this morning is not very good. It's snowing and there's a light wind blowing. Enjoy the rest of your flight. Thank you for travelling with us today.

Lesson 15 Grammar, activity 2

Conversation 1

MAN Everything all right, darling?
WOMAN Yes, everything's fine. The food is delicious.
MAN Good, I'm delighted. I'm very fond of this restaurant. The chef is excellent.
WOMAN Do you come here often?
MAN Ah, well, only with... with very special people.
WOMAN Such as?
MAN Well, such as business clients...

Conversation 2

WOMAN Look at those oranges! Let's have some, shall we?
MAN 1 But we've got a lot of fruit – bananas, apples, melons. We don't need oranges as well.
WOMAN And we can get some potatoes and eggs here. How much are the potatoes?
MAN 2 Thirty pence a kilo.
WOMAN OK, I'd like two kilos.
MAN 1 Two kilos of potatoes.
MAN 2 Two kilos of potatoes, there you go.

Conversation 3

MAN What's happened?
WOMAN I don't know.
MAN Excuse me, what's happening?
DRIVER I don't know.
MAN Well, how long are we staying here?
DRIVER I don't know.
MAN But I'm going to work. I'm late already. What am I going to do?
DRIVER I don't know.
MAN Do you know why you're beginning to annoy me?
DRIVER No.
MAN You keep saying 'I don't know.'

Progress check 11 to 15 Vocabulary, activity 5

Conversation 1

MAN Excuse me!
WOMAN Yes, sir. Can I help you?
MAN Yes, I'm looking for the men's department.
WOMAN It's over there!
MAN Thank you.

Conversation 2

MAN Can you carry this for me, please?
WOMAN Pardon? I didn't hear what you said.
MAN I said can you carry this for me, please?
WOMAN Yes, of course.

Conversation 3

MAN Oh, sorry!
WOMAN That's OK.

Lesson 16 Speaking and listening, activity 2

SPEAKER 1 Her name was Mrs Smith. She was a tall lady, and was very old, or so it seemed. But she was a very good teacher, and she was very kind to us.
SPEAKER 2 All my friends were at my party. I remember it was a sunny day, in June, and we were all in the garden, playing together. There was lots to eat and drink and lots of games to play. Then they all sang Happy Birthday.
SPEAKER 3 There was a large map of the world on the walls of the classroom, on the one side of the board, and on the other, there was a chart to show us how to write the alphabet.
SPEAKER 4 His name was Jack, and we were very good friends. Our mothers were very good friends too, so I saw Jack nearly every day for the first three or four years of my life. Then we were at different schools, so we weren't together so often.

Lesson 17 Vocabulary and listening, activity 4

JEAN OK, what do we need?
TONY We need some fruit and vegetables.
JEAN How about some oranges?
TONY OK, and we'll have some bananas.
JEAN Yes, there aren't any bananas. And let's get some apples.
TONY OK, apples. And we haven't got any onions.
JEAN A kilo of onions. That's enough. And some carrots?
TONY That's right, we haven't got any carrots. And let's get some meat.
JEAN Yes, OK. You like chicken, don't you?
TONY Yes, chicken's great. And we need some tomatoes.
JEAN OK, two kilos of tomatoes. Anything else?
TONY No. Oh, have we got any water?
JEAN No, we need a couple of litres of water and let's get some juice. That's it.

Lesson 17 Listening and speaking, activity 2

Q Is it true, Lisa, that you always have bacon and eggs for breakfast?
LISA Well, it used to be true, but it isn't true any more. People often have toast and cereal, jam, yoghurt, things like that, but not many have time to cook bacon and eggs. It's only in hotels when you get bacon and eggs – what we call a cooked breakfast, or an English breakfast.
Q Do you always have meat and vegetables at the main meal?
LISA Not always, no. In any case, I don't eat meat.
Q Do people in Britain drink a lot of wine?
LISA Yes, they do, but not at every meal. They don't drink wine at lunch and dinner every day. Perhaps we'll have a glass of wine at the weekend.
Q What do people drink then?
LISA Water, juice. Some families drink tea with their meals.
Q Do you drink *much* tea?
LISA Yes, people drink tea during the day. Many people will have five or six cups of tea a day. We usually have it with milk. Other people prefer coffee.
Q What's the main vegetable that you find at most meals?
LISA I suppose we often eat potatoes with our main meal. Potatoes are very popular. But we also eat a lot of pasta, but we don't eat pasta *and* potatoes!
Q Are there many people who don't eat meat in Britain?
LISA Yes, there are many vegetarians like me in Britain. Every restaurant will always offer several vegetarian dishes, and when people come to dinner you always check to see if there are any vegetarians.

Lesson 18 Listening and speaking, activity 1

MAN Whitney Houston was born in New Jersey in 1963. She started singing when she was eleven years old. As a teenager she worked with rhythm and blues singers such as Chaka Khan and Lou Rawls. In 1985 she had a hit with her first album, *Whitney Houston*. She received a Grammy Award with the song *Saving all my love for you*. In 1986 she was the first pop singer to sell ten million copies of her first two albums. In 1992 she appeared with Kevin Costner in the film *The Bodyguard*. The following year she had another hit with *I will always love you* from the film. She isn't married.

Lesson 19 Functions, activity 1

Conversation 1

MAN I've heard so much about your mother. Tell me about her. What's she like?
WOMAN Well, she's quite elderly now. She's medium-height, with short hair. She's still quite attractive, I think. What else can I say? Oh, well, you can see for yourself in an hour.
MAN I am looking forward to meeting her.

Conversation 2

WOMAN 1 So you've got a new boyfriend, I hear.
WOMAN 2 That's right.
WOMAN 1 What's he like, then.
WOMAN 2 Well, he's middle-aged and rather good-looking.
WOMAN 1 What does he look like?
WOMAN 2 Oh, he's quite tall, with dark hair.
WOMAN 1 And what does he do?
WOMAN 2 He's an accountant.
WOMAN 1 Oh.

Conversation 3

MAN And you say he's going to arrive by plane from Frankfurt.
WOMAN Yes, that's right.
MAN How will I recognise him? What's he like?
WOMAN Well, he's short and he's got dark curly hair. He's quite well-built.
MAN Anything else?
WOMAN Oh, and he's got a moustache.
MAN I see. How old is he?
WOMAN He's twenty.
MAN I suppose I should write his name on a card and hold it up as the passengers come out of the arrivals lounge.
WOMAN Oh, I'm sure you'll recognise him.

Conversation 4

MAN So, what's the new woman at work like?
WOMAN Oh, she seems very nice. She's quite young. About thirty. She's tall and slim with long fair hair. And she wears glasses.
MAN She sounds very nice. When do I meet her?
WOMAN Oh, she's not your type, John.
MAN Not my type? Tall? Slim? Long fair hair? Not my type?
WOMAN No, John.
MAN Why not?
WOMAN Well, have a look in the mirror, John.
MAN Some people think I'm very good-looking.

Lesson 19 Listening and speaking, activity 2

Q Would you say that the Irish are tall people, Kevin?
KEVIN No, they're not very tall, I suppose we think someone is quite tall when they're over one metre eighty, something like that.
Q And when does old age begin, in your opinion? How old are people when they're old?
KEVIN I suppose about sixty or seventy. No, it used to be sixty, but it's changing. No, I'd say seventy.
Q And middle age? When are people middle-aged?
KEVIN I'd say you're middle-aged when you're over forty.
Q And would you say that people are well-built or slim?
KEVIN Oh, it's hard to say. I suppose some of us are well-built, some of us are slim. Yes, I think it is difficult to say.

Lesson 20 Vocabulary and listening, activity 3

MAN Hi, Mary, Hi, Bill!
BILL Hello, how are you?
MAN Fine thanks, how was your trip?
MARY It was great. We had a wonderful time.
MAN Where did you go?
MARY Well, we flew to Paris, where we did everything the tourists usually do. We walked a lot by the river...
BILL And the weather was sunny...
MAN Great!
MARY And we visited the museums and galleries, and the Eiffel Tower...
BILL And did some shopping.
MARY Yes, we did some shopping.
BILL No, *you* did some shopping.
MARY OK, OK, I did a lot of shopping.
BILL Then we went to Venice, where we met some New Yorkers, and we did a lot of sightseeing with them.
MARY We had a nice time together.
BILL Then we went to Vienna...
MARY I've always wanted to go to Vienna.
BILL And we found a cheap hotel and what did we do there?
MARY We went to the Opera. It was wonderful. So romantic. And then we went to Budapest.
BILL Yeah, and Mary bought some souvenirs, while I had a steam bath. It was great.
MARY Except you lost your wallet.
BILL Yes, I lost my wallet, but someone found it and luckily it had the name of my hotel in it and I got it back.
MARY Isn't that great?
MAN Amazing.
MARY And then we flew back to Paris, and then flew home.
MAN Well, welcome home!
BILL & MARY Thank you.

Progress check 16 to 20 Sounds, activity 4

WOMAN Was Jonas born in Vienna?
MAN Was there anything to eat at the party?
WOMAN Did you see the football match last night?
MAN Were there lots of people on the bus?
WOMAN Did they arrive on time?
MAN Did she get home safely?

Lesson 21 Sounds

What did Agatha Christie write?
When was she born?
Where did she live?
Who did she marry?
When did she disappear?
Where did her husband find her?
When did he find her?
What did Sir Max Mallowan do?

Lesson 22 Listening and speaking, activity 2

SPEAKER 1 Oh, it was five months ago, in August, we were talking about it yesterday evening. It was a typical wedding day, which started with the excitement of getting dressed in my wedding dress. Then we drove to the church, and we arrived about five minutes late, which is the custom. Then the service started and after the service, there was a photographer. And finally we went to the wedding reception, which was wonderful, and we danced until about three in the morning. It was a wonderful day.
SPEAKER 2 It was last Thursday when I heard my exam results. After three years of hard work at university, at the end of the year, the university year, that is, finally the day arrived when I found out if I had passed the exam. I woke up early and went down to the examination building where they put up the exam results on lists. The building was closed so I waited about ten minutes with a number of other students. Then they opened the doors and we went in, and I found my name on the list. I'd passed! What a

great feeling! I left the building, dancing and singing, people must have thought I was mad.
SPEAKER 3 It was in 1987, on the eleventh of December, that my parents had their golden wedding anniversary, that's fifty years of marriage. And we had a big party for them, all my brothers and sisters, all the grandchildren, nephews, nieces, friends and neighbours, there were about sixty people in all. We worked all day from nine to five preparing the food and decorating the house. And the funny thing was – they didn't know about the party until they got home. They walked in the door and everyone started singing. It was a great surprise!

Lesson 23 Vocabulary and listening, activity 5

MAN Hey, this is a great party.
WOMAN I'm glad you were able to come.
MAN You must introduce me to your friends. Which one is Harriet? I've heard so much about her.
WOMAN Well, can you see the woman by the door?
MAN Er, which one?
WOMAN The one in the jeans.
MAN No, which door?
WOMAN There is only one door.
MAN Oh, right! Yes, I see. The one in the jeans and the T-shirt.
WOMAN Yes, she's standing by the door and talking to the middle-aged man. She's smiling at him.
MAN Great smile. And what about John?
WOMAN John is sitting down in the armchair by the window.
MAN Is he the one wearing the sweater and trousers?
WOMAN No, John's wearing the shirt and tie.
MAN OK, I see him. Nice tie. Love that yellow.
WOMAN And Edward is the man over by the window. He's laughing at something. Probably laughing at John's tie.
MAN The man in the trainers with the blue shirt and black trousers?
WOMAN That's Edward!
MAN Right! And Louise?
WOMAN Louise is standing by the television. She's wearing a black dress.
MAN Yes, I can see her. Hey, she's smiling at me.
WOMAN She smiles at everyone. She wears glasses, but she hasn't got them on at the moment.
MAN No, I'm sure she's smiling at me...

Lesson 24 Reading and listening, activity 3

Speaker 1

MAN Well, we find that our jobs take up a lot of time, and when the weekend comes, we're very tired.
WOMAN But the trouble is, we don't have any social life now.
MAN So our New Year's Resolution is we're going to invite more friends for dinner.
WOMAN Yes, because we don't entertain much. And we'd like people to invite us as well.

Speaker 2

MAN I've got a year off before I go to university, and I don't know much about foreign countries, so I'm going to travel around Europe. I always stay in Britain for my holiday, and I'd like to see how our neighbours live.

Speaker 3

WOMAN I spend my life driving to school, teaching and then coming home. So I'm going to get fit because I don't take enough exercise. I'm going to start running in the evening when I get home.

Speaker 4

MAN We've got a family now, and we need more space for all the things that young children need. Our resolution is that we're going to move.
WOMAN Yes, because our house is too small for four people. The trouble is, we still like it here, though.
MAN Yes, it's going to be hard to leave this house.

Lesson 25 Vocabulary and listening, activity 5

ASSISTANT Good afternoon. Can I help you?
CUSTOMER Good afternoon. Yes, I'd like a burger with fries and a Coke, please.
ASSISTANT Would you like a regular or a large Coke?
CUSTOMER A regular, please.
ASSISTANT Would you like anything else?
CUSTOMER Yes, I'd like an ice cream, please.
ASSISTANT What flavour would you like?
CUSTOMER Strawberry, please.
ASSISTANT OK.
CUSTOMER How much is that?
ASSISTANT That's four pounds fifty.
CUSTOMER Here you are.
ASSISTANT Thank you.

Lesson 26 Speaking and listening, activity 4

ASSISTANT Can I help you?
CUSTOMER Yes, I'm looking for a sweater.
ASSISTANT We've got some sweaters over here. What colour are you looking for?
CUSTOMER This blue one is nice.
ASSISTANT Yes, it is. Is it for yourself?
CUSTOMER Yes, it is. Can I try it on?
ASSISTANT Yes, go ahead.
CUSTOMER No, it's too small. It doesn't fit me. Have you got one in a bigger size?
ASSISTANT No, I'm afraid not. What about the red one?
CUSTOMER No, I don't like the colour. Red doesn't suit me. OK, I'll leave it. Thank you.
ASSISTANT Goodbye.

Lesson 26 Vocabulary and listening, activity 4

Conversation 1

ASSISTANT Can I help you, sir?
CUSTOMER Yes, I'm looking for something for my wife. Have you got any perfume?
ASSISTANT Yes, sir. Try this. It's our new perfume for this season. It's called *Reward*.
CUSTOMER Can I try it? Mm. It's very good. How much is it?
ASSISTANT This bottle costs £37.
CUSTOMER Hm. Have you got it in a smaller bottle?

Conversation 2

CUSTOMER Excuse me.
ASSISTANT Yes, madam, can I help you?
CUSTOMER I'd like a box of chocolates, for my friend.
ASSISTANT Yes, madam. 500 grams?
CUSTOMER How much is that?
ASSISTANT Ten pounds. These are particularly good chocolates.
CUSTOMER OK, I'll have them.

Lesson 27 Listening and speaking, activity 2

OFFICIAL Good afternoon, how can I help you?
CUSTOMER I've lost my bag. I put it down somewhere in the market, and forgot about it.
OFFICIAL Well, don't worry, madam, it'll turn up. I'll just take down some details. Could you tell me your name, please?
CUSTOMER Yes, my name's Jill Fairfield.
OFFICIAL Jill Fairfield. Is that Mrs Fairfield?
CUSTOMER Ms Fairfield.
OFFICIAL OK, and your address?
CUSTOMER 32, Burn Road, Manchester.
OFFICIAL Burn Road, Manchester. OK, and your telephone number?
CUSTOMER 679 5453.
OFFICIAL Right and you say it was a bag that you lost?

CUSTOMER That's right.
OFFICIAL And you lost it in Chester market?
CUSTOMER That's correct. At about ten in the morning.
OFFICIAL And that was today, right?
CUSTOMER Yes.
OFFICIAL So, Thursday the twenty-first of July. And can you describe it to me?
CUSTOMER Well, it was large, square and it was made of black nylon.
OFFICIAL And was there anything in it?
CUSTOMER Yes, there was my purse, a calculator, an address book, a newspaper and a comb.
OFFICIAL Right, now, let me see. Was it this one?
CUSTOMER No, that's not mine.
OFFICIAL Is this yours?
CUSTOMER Yes, that's mine. Thank you very much.

Lesson 27 Listening and speaking, activity 5

Conversation 1

CUSTOMER Excuse me!
OFFICIAL Yes, sir, good morning.
CUSTOMER I'd like to report the loss of a coat.
OFFICIAL Yes sir. And your name is...?
CUSTOMER Ken Hamilton.
OFFICIAL Ken Hamilton, all right. And your address?
CUSTOMER 13, Dock Lane, London.
OFFICIAL And your phone number?
CUSTOMER 75859.
OFFICIAL And when did you lose it?
CUSTOMER On the thirty-first of May around four in the afternoon, I must have left it on the train.
OFFICIAL And what was it like?
CUSTOMER It was made of red leather.
OFFICIAL Red leather! Was it... a man's coat, sir?
CUSTOMER Er, well, yes... er...
OFFICIAL I see. Now, did you lose anything else?

Conversation 2

CUSTOMER And I must have lost it then.
OFFICIAL Just say your name again, madam.
CUSTOMER Mary Walter.
OFFICIAL And your address and phone number?
CUSTOMER 21, Tree Road, Leeds, 75889.
OFFICIAL And it was a black plastic bag, you say?
CUSTOMER Yes.
OFFICIAL And you last saw it on the twentieth of March at two in the afternoon?
CUSTOMER Yes, in the supermarket.
OFFICIAL And what was in it?
CUSTOMER All my shopping and my purse.

Conversation 3

CUSTOMER J-O-S-E-P-H.
OFFICIAL OK, Mr Joseph. And the address?
CUSTOMER Where I left it?
OFFICIAL No, your address.
CUSTOMER 33, James Street, Bath.
OFFICIAL Do you have a daytime telephone number?
CUSTOMER Yes, 56778.
OFFICIAL What was this box of cigars made of?
CUSTOMER Well, it was wooden and square.
OFFICIAL I see. And you left it on the bus?
CUSTOMER Yes, yesterday morning at about eleven o'clock.
OFFICIAL OK, about eleven o'clock on the seventeenth of October.

Lesson 28 Vocabulary and listening, activity 2

Conversation 1

DOCTOR Good morning. How are you?
PATIENT Fine, thanks.
DOCTOR So, if you're fine, why are you here to see me?
PATIENT No, what I meant was, oh, it doesn't matter. I've got a headache. I seem to have it all the time.
DOCTOR I see. Any other symptoms?
PATIENT Well, I've got a cough as well.
DOCTOR Do you smoke?
PATIENT Yes I do. And I feel tired all the time.
DOCTOR OK, let's have a look.

Conversation 2

DOCTOR And what seems to be the matter with you?
PATIENT I feel sick and I've got a stomachache.
DOCTOR Let me see. Have you got a headache?
PATIENT Yes, I have.
DOCTOR You look rather hot. Yes, you've got a bit of a temperature. I think it must be something you ate yesterday.
PATIENT I only had a sandwich yesterday.
DOCTOR What kind of sandwich?
PATIENT It was a cheese sandwich.
DOCTOR Well, it's probably nothing serious, but I'll give you some medicine...

Conversation 3

DOCTOR And what seems to be the trouble?
PATIENT I've hurt my leg.
DOCTOR How did you do that?
PATIENT In a game of football.
DOCTOR Football! Don't you think you're too old to play football?
PATIENT Well, I'm only seventy-three.
DOCTOR Really! Well, let me see now...

Lesson 29 Reading and listening, activity 2

Q Karl, tell me something about Sweden. Is it a large country?
KARL Yes, it is quite large, at least for a country in Europe. It's almost 450,000 square kilometres.
Q 450,000. Yes, that's quite big.
KARL Yes, it's nearly twice the size of Britain.
Q I see. And what's the coldest month? January, I suppose?
KARL Yes, January is quite cold. The temperature is about minus three degrees Celsius.
Q And how hot is it in the summer?
KARL Actually, it gets quite warm. In July, the average temperature is eighteen degrees.
Q And what's the average rainfall?
KARL Well, for Sweden, it's 535mm, but it's less in Stockholm.
Q And what's the population?
KARL There are over eight million people. Not so many for such a large country.
Q And what about the armed services? How many troops are there?
KARL We only have about 65,000 soldiers.
Q 65,000, I see. And when do children start school?
KARL They start at the age of seven and continue for ten years, until they're seventeen.
Q Ten years.
KARL Of course, some students go on to university.

Lesson 30 Vocabulary and listening, activity 5

KATY What's your favourite sport, Andrew?
ANDREW Well, I like most sports, but I suppose I like football most of all. Like most people.
KATY Yes, I suppose football is the most popular sport. Personally, I don't like football. I don't enjoy competitive sports. I like cycling and horseriding.
ANDREW Isn't horseriding very expensive?

KATY Yes, it's more expensive than cycling.
ANDREW I think horseriding is the most expensive sport. What do you think is the most tiring sport?
KATY Well, horseriding is very tiring.
ANDREW Do you think it's more tiring than, say, tennis?
KATY Oh, yes, I'm exhausted after I've been horseriding. What about you?
ANDREW Well, for me tennis is the most tiring. What do you think is the most dangerous sport?
KATY I think hanggliding is very dangerous.
ANDREW Well, that's what many people think. But you know, there are more accidents to do with windsurfing than there are with hanggliding.
KATY I didn't know that. Which is the most difficult sport, in your opinion?
ANDREW How about climbing? I think climbing is very hard.
KATY Well, I think skiing is more difficult than climbing.
ANDREW No, I don't agree. Climbing looks incredibly difficult.
KATY And what do you think is the most exciting sport?
ANDREW Well, tennis, I think. What about you?
KATY It has to be motor racing. Motor racing is the most exciting sport for me.

Lesson 31 Listening, activity 1

JAMES Well, in Australia, you needn't ask if you want to take a photograph of someone you don't know. Even someone you do know will not usually mind if you take a photo.
Yes, I've heard about this custom with shoes in Japan. Actually, it's quite a good idea. But you needn't take your shoes off when you go into someone's house in Australia, unless of course, they're very dirty. I usually change my shoes when I get home, but I don't take them off when I visit people.
And women needn't cover their heads in Australia, most of the time. Sometimes, in certain churches, women wear a hat or a scarf on their heads, but they needn't do so in the street or at work.
Pointing at people, yes, that's true in Australia, you mustn't point at people. If you do, people think you're rather rude. It's the same in Britain, I think.
In Australia you can look people in the eye, though. It shows you're interested in them and what they're saying. It's a sign of politeness. But you mustn't look people in the eye for long, as they begin to feel uncomfortable.
Well, you *can* kiss in public in Australia, but not many people do so. We're fairly relaxed in my country, so almost anything is OK, but, well... let's say it's only young people who kiss in public – it's not forbidden, but it's not very common.
If you give a gift, you needn't use both hands. You can use just one hand. There aren't any rules about this sort of thing.
And, no you needn't shake hands with everyone when you meet them in Australia. You can shake hands when you meet someone for the first time, in fact, it's bad manners if you don't, but not every time, no.

Lesson 31 Sounds, activity 1

Children mustn't play near the road.
You must be quiet in a library.
You must keep your wallet in a safe place.
Men must take off their hats in a church.
You mustn't give a gift with one hand in Taiwan.
You mustn't wear shoes in a Japanese home.

Lesson 33 Listening and vocabulary, activity 3

ALAN Hi! How's your day been?
BARRY Awful, absolutely awful.
ALAN I'm sorry to hear that. What's happened?
BARRY Well, I've hurt my back.
ALAN Your back! How did you hurt it?
BARRY I tried to lift a box of books.
ALAN A box of books! I'm not surprised you hurt yourself trying to lift a box of books. Have you been to the doctor yet?
BARRY No, not yet.
ALAN Well, I think you should go immediately. And what else has happened?
BARRY I've lost my wallet.
ALAN Your wallet? Where did you lose it?
BARRY At the bus stop, I think.
ALAN Have you been back to the bus stop yet?
BARRY No, I haven't.
ALAN And have you heard your exam result?
BARRY Yes, I have.
ALAN Have you passed?
BARRY No, I've failed it.
ALAN Oh dear, it's been one of those days for you, hasn't it?

Lesson 33 Listening and vocabulary, activity 6

Conversation 1

WOMAN Have you seen my bag?
MAN No, where did you leave it?
WOMAN Right here under my desk. But it's not there now.
MAN When did you last see it?
WOMAN This morning.
MAN Oh no! Are you thinking what I'm thinking.
WOMAN I think someone has stolen it.
MAN That's right. I think someone has stolen your bag.

Conversation 2

MAN Come on!
WOMAN I can't run any faster.
MAN It's not far. We can make it.
GUARD Sorry, sir, it's too late.
MAN What do you mean... too late?
WOMAN We've got to catch the Waterloo train.
GUARD I'm sorry but you've just missed the Waterloo train. It's just left.
WOMAN Oh no! Now what are we going to do?

Conversation 3

MAN And Williams comes along the right wing and passes to Franks and Franks runs with the ball, round Gray, and flips it over the goalkeeper's head and he's scored, he's scored a goal for United. What a brilliant goal!

Conversation 4

WOMAN Aargh!
MAN What's the matter! What have you done!
WOMAN I've cut my finger.
MAN You've cut your finger! Oh, I am sorry. Don't worry. I know exactly what to do.
WOMAN What?
MAN I should put my head between my knees so I don't faint at the sight of blood.
WOMAN Thanks. Thanks very much for your help.

Lesson 35 Vocabulary and sounds, activity 2

Conversation 1

MAN Are you asleep?
WOMAN No, not yet. But I'm very tired.
MAN Did you hear that?
WOMAN Hear what?
MAN That noise.
WOMAN No, I didn't.
MAN I think I'll find out what it is.

Conversation 2

MAN Could you move your car, please?
MAN 2 What?
MAN I said, could you move your car?
MAN 2 Why should I?
MAN Because it's in my way.

Conversation 3

WOMAN I'm so sorry to trouble you, but could you tell me how to get to Thames Street?
MAN Thames Street. Well, if I were you, I wouldn't start from here.

Conversation 4

TEACHER Quiet! ... I said quiet! Stop shouting, Smith! Sit down, White! Get your books out! Will you all be quiet!

Lesson 35 Listening and speaking, activity 2

JENNY Look, there are these questions here about how you got on at school. Shall we, shall we just go through them?
GAVIN Yes, let's.
JENNY OK, so, did you always work very hard?
GAVIN Er, well I certainly worked pretty hard at the subjects I enjoyed, yes I did. What about you?
JENNY Yes, I did actually, I think I worked very hard, yeah. And what about this other one?
GAVIN Did, yeah, did you always listen carefully to your teachers?
JENNY Erm... no I don't think I did. No, I think I was quite disruptive, actually. What about you?
GAVIN Well, I think I did listen to the teachers certainly when I got to the level where I was doing the subjects that I enjoyed.
JENNY Yeah, OK, this next one says, did you always behave well?
GAVIN I don't think I did always behave that well. I was erm, a bit, er, a bit of a tearaway.
JENNY Mm. Well, I think I was pretty well-behaved on the whole, so I'd say yes, yeah.
GAVIN Good for you! Did you pass your exams easily?
JENNY Erm, no I can't say I did, no, I, I found them quite a struggle, actually. What about you?
GAVIN I didn't pass them that easily, because I worked hard but also I found it very difficult writing all that amount of material in such a short amount of time.
JENNY Yeah, yeah, exactly. What about this one, then? Did you always write slowly and carefully?
GAVIN Quite slowly. Essays took a long time to write and I suppose I took quite a bit of care, yes.
JENNY Yes, I agree. I was also, I was very careful and erm, yeah, yeah I was quite methodical.
GAVIN And did you think your school days were the best days of your life?
JENNY Um, no, no I can't say they were. What about you?
GAVIN No, I went away to boarding school when I was quite young and I didn't like that. No, I, they weren't the best days of my life.

Lesson 36 Listening, activity 2

OFFICER Can I help you?
CUSTOMER Yes, I'd like a ticket to Birmingham.
OFFICER When do you want to travel? It's cheaper after 9.30.
CUSTOMER I'll travel after 9.30.
OFFICER Single or return?
CUSTOMER I'll have a single ticket, please.
OFFICER That'll be thirty pounds exactly. How would you like to pay?
CUSTOMER Do you accept credit cards?
OFFICER I'm afraid not.
CUSTOMER Well, I'll pay cash, then. Will there be refreshments on the train.
OFFICER Yes, there will.
CUSTOMER Can I have a ticket for the car park as well?
OFFICER That'll be thirty-two pounds in all.
CUSTOMER Thank you.

Lesson 37 Vocabulary and listening, activity 4

FORECASTER And here's the weather forecast for the rest of the world. Athens, cloudy twelve degrees. Bangkok, cloudy thirty degrees. Cairo sunny sixteen degrees. Geneva, cloudy ten degrees. Hong Kong, cloudy twenty degrees. Istanbul, rainy seven degrees. Kuala Lumpur, sunny thirty-five degrees. Lisbon, cloudy eleven degrees. Madrid, rainy seven degrees. Moscow, snowy minus ten degrees. New York, sunny zero degrees. Paris, snowy minus six. Prague, sunny minus two. Rio, cloudy minus 29. Rome, rainy nine degrees. Tokyo, snowy minus four degrees and finally Warsaw, cloudy minus eight degrees.

Lesson 37 Reading and listening, activity 4

Q And we have a scientist with us today to discuss various predictions about our weather in the future. Professor Stein, what do you think about the latest report about our climate?
STEIN I think the predictions are, in general, very accurate.
Q So you think temperatures will rise in the future?
STEIN Yes, in the next twenty-five years, they'll rise by two to six degrees.
Q And what will the consequences of that be?
STEIN Well, the ice at the North and South poles will melt and the sea level will rise.
Q Do you think whole countries will disappear?
STEIN No, I don't. That won't happen for another hundred or more years.
Q And will there be enough fresh water for everyone?
STEIN Yes, there will. But it won't come from rainfall, which will decrease in general, but from the sea with the salt taken out.
Q Will fresh water cost more?
STEIN Yes, certainly. And this will mean that factory goods will cost more to produce.
Q What effect will this have on the economy? Will it get worse?
STEIN No, I don't think so. But we'll have to change our lifestyles in the future.

Lesson 38 Listening, activity 1

FRANK So, Sally, this quiz then, shall we have a go? Right, number one, coffee is grown in a, Brazil, b, England, c, Sweden. What do you think?
SALLY Oh, well, that's got to be Brazil hasn't it?
FRANK I think so, yeah, that's, a.
SALLY OK, number two, Daewoo cars are made in Switzerland, Thailand or Korea?
FRANK Korea, definitely.
SALLY Yeah? OK, so that's, c.
FRANK Number three, Sony computers are made in Japan, USA or Germany?
SALLY Japan.
FRANK Mm... a, then.
SALLY OK, erm, number four. Tea is grown in a, India, b, France or c, Canada?
FRANK I think it must be India, mustn't it. Don't you think?
SALLY Yeah, definitely, so that's, a for that one.
FRANK OK, number five, tobacco, where's tobacco grown then, Norway, Iceland or the USA? Well it's too cold for Iceland.
SALLY Yeah, it's the USA.
FRANK The USA, that's right, c then.
SALLY OK, Benetton clothes are made in Italy, France or Malaysia?
FRANK Benetton, I think, that's Italy, don't you?
SALLY Yeah, I think it is. Yeah.
FRANK a.
SALLY a. ... er, number seven, Roquefort cheese is made in a, Germany, b, Thailand or c, France?
FRANK Erm.
SALLY It's not Germany, is it?
FRANK No, I don't think it's Germany, I think Roquefort is France.
SALLY Yeah, France, OK.
FRANK Right, number eight. The atom bomb was invented by the Japanese, the Americans or the Chinese?
SALLY The Americans.
FRANK Yeah, the Americans, b.
SALLY Er, *Guernica* was painted by Picasso, Turner or Monet?
FRANK Um, that's Picasso, I think, that one.
SALLY Is it?
FRANK Yeah, yeah, famous painting that, a.
SALLY Right, the West Indies were discovered by Scott of the Antarctic,

Christopher Columbus or Marco Polo?
FRANK I think that's Christopher Columbus, was the one.
SALLY Was it, are you sure?
FRANK Yeah, I think, pretty sure.
SALLY OK, that's b.
FRANK OK, telephone, who invented the telephone, Bell, Marconi or Baird? Baird invented the television, I think.
SALLY Oh, it's Bell.
FRANK Bell?
SALLY Yeah, definitely.
FRANK OK, twelve, *Romeo and Juliet*, who wrote *Romeo and Juliet*? That's simple, isn't it?
SALLY Yeah.
FRANK Go on, then.
SALLY Well, it wasn't Ibsen.
FRANK No, it's got to be Shakespeare.
SALLY Yeah, and it wasn't Primo Levi.
FRANK No, that's, b, OK.
SALLY Number thirteen. The Blue Mosque in Istanbul was built by a, Sultan Ahmet I, b, Ataturk or c, Suleyman the Magnificent?
FRANK Well, I know this one.
SALLY Do you?
FRANK Yup, it was Ataturk.
SALLY No, Sultan Ahmet I.
FRANK Oh. *Yesterday*, right who composed *Yesterday*, Paul McCartney, John Lennon, well, it certainly wasn't Mick Jagger.
SALLY No.
FRANK So Paul McCartney or John Lennon? I think I know.
SALLY I think it's, a, Paul McCartney
FRANK Mm. And finally, the pyramids, who were they built by, the Pharaohs, the sultans or the council? Wasn't my local council!
SALLY No, erm...
FRANK I think it was, it's...
SALLY It's the Pharaohs isn't it? Yeah, that's, a.

Lesson 39 Listening and reading, activity 2

CHRIS Good afternoon.
RECEPTIONIST Good afternoon. Can I help you?
CHRIS Have you got any beds for tonight?
RECEPTIONIST Yes, I think so. Sorry, but I've just started work at the hostel. How long would you like to stay?
CHRIS We'll stay for just one night.
RECEPTIONIST Yes, that's OK.
TONY Great!
RECEPTIONIST How old are you?
TONY We're both sixteen.
RECEPTIONIST One night's stay costs £6.50 each.
CHRIS Is it far from the hostel to the centre of Canbury.
RECEPTIONIST Yes, it's two kilometres. It takes an hour on foot.
TONY Is there a bus service?
RECEPTIONIST I think so. It takes about fifteen minutes. There's a bus every hour.
TONY When does the last bus leave the city centre?
RECEPTIONIST I think it leaves at nine o'clock in the evening. There's not much to do in the evening.
CHRIS We're very tired. We need an early night. What time does the hostel close in the morning?
RECEPTIONIST Er, at eleven am. Where are you walking to?
CHRIS We're going to Oxton. Are you serving dinner tonight?
RECEPTIONIST Yes, we're serving dinner until eight o'clock. And breakfast starts at seven-thirty.
TONY And where's the next hostel?
RECEPTIONIST I'm not sure. I think it's Kingscombe, which is about ten kilometres away. I started work last Monday so I'm very new here.

Lesson 39 Listening and reading, activity 4

CHRIS It's very strange. She said one night's stay cost £6.50, but it costs £6.15.
TONY Yes, and she said it was two kilometres to the city centre.
CHRIS But, in fact, it's three kilometres.
TONY And she said the last bus left at nine o'clock. But it leaves at eight o'clock.

Lesson 39 Grammar, activity 3

CHRIS And she said the bus took fifteen minutes. But in fact, it takes ten minutes.
TONY And she said the hostel closed at eleven am, but it's open all day.
CHRIS It seems that they serve dinner from six to seven.
TONY But she said they were serving until eight o'clock. And she also said breakfast started at seven-thirty...
CHRIS ... when, in fact, it says here that breakfast starts at seven.
TONY And she said that Kingscombe was the next hostel, but it isn't. It's Charlestown.
CHRIS And finally she said that Kingscombe was ten kilometres away. But it's fifteen kilometres.

Lesson 40 Reading and listening, activity 6

Ruth's Parents

The next afternoon, Jan went to Ruth's house for tea.
'How do you do, Mr and Mrs Clark,' Jan said.
'Sit down, Jan,' said Mrs Clark. 'Would you like a cup of tea?'
'Yes, please,' said Jan. He didn't feel very comfortable.
'With sugar?' said Mrs Clark.
'Yes, please,' Jan answered. 'Sugar, but no milk.'
'No milk?' said Mrs Clark. 'But everybody drinks milk in tea!'
'Oh...' said Jan. 'Well, in Poland, we never have milk in our tea.'
Mr and Mrs Clark looked at Jan. 'These foreigners have strange ideas,' said Mr Clark.
Jan stayed for about an hour. Mr Clark spoke English very quickly and Jan did not always understand.
Outside the door, Jan said to Ruth, 'Your parents don't like me very much.'
'Don't be silly, Jan,' said Ruth. 'My parents haven't met many foreigners. It's all right.'
'OK, Ruth,' said Jan. 'I'll see you tomorrow.' And he walked away. But he felt unhappy.
Later that evening Ruth asked, 'Well, Mum, did you like Jan?'
Ruth's mother said, 'Well, he didn't speak English very well. Your father and I liked Bill. What's wrong with an English boyfriend? And Jan is going back to Poland soon.'
'But I don't like Bill any more,' shouted Ruth and ran out of the room. 'I don't like Bill,' she said to herself, 'but I do like Jan. Maybe I love him.'

Wordlist

The first number after each word shows the lesson in which the word first appears in the vocabulary box. The numbers in *italics* show the later lessons in which the word appears again

address /əˈdres/ 3
adult /ˈædʌlt/ 39
afternoon /ˌɑːftəˈnuːn/ 7
age /eɪdʒ/ 3
airport /ˈeəpɔːt/ 12
album /ˈælbəm/ 16-20
appear /əˈpɪə(r)/ 18
apple /ˈæp(ə)l/ 17
apple pie /ˈæpe(ə)l paɪ/ 25
April /ˈeɪprɪl/ 22
arm /ɑːm/ 28, *33*
armchair /ɑːmˈtʃeə(r)/ 8
arrival hall /əˈraɪv(ə)l hɔːl/ 36
artist /ˈɑːtɪst/ 2
aspirin /ˈæsprɪn/ 28
at the back /æt ðə bæk/ 8
at the front /æt ðə frʌnt/ 8
attractive /əˈtræktɪv/ 15, *16-20, 19*
August /ˈɔːgəst/ 22
aunt /ɑˈnt/ 6
Australia /ɒsˈtreɪlɪə/ 1
Austria /ˈɒstrɪə/ 20

back /bæk/ 28
bacon /ˈbeɪkən/ 17
badly /ˈbædlɪ/ 35
bad-tempered
/bæd ˈtempəd/ 16
bag /bæg/ 5, *33*
baggage reclaim
/ˈbægɪdʒ rɪˈkleɪm/ 36
baker /ˈbeɪkə(r)/ 14
ballet /ˈbæleɪ/ 11-15
banana /bəˈnɑːnə/ 17
Bangkok /ˈbæŋgkɒk/ 1
bank /bæŋk/ 14
bar /bɑː(r)/ 26
barbecue /ˈbɑːbɪˌkjuː/ 34
basketball /ˈbɑːskɪtˌbɔːl/ 30
bath /bɑːθ/ 8
bathroom /ˈbɑːθruːm/ 8
beard /ˈbɪəd/ 19
beautiful /ˈbjuːtɪˌfʊl/ 4, *15*
bed /bed/ 8
bedroom /ˈbedruːm/ 8
beef /biːf/ 17
beer /bɪə(r)/ 25
bicycle /ˈbaɪsɪk(ə)l/ 12
big /bɪg/ 15, *29*
bill /bɪl/ 33
biscuits /ˈbɪskɪtz/ 26
black /blæk/ 5
blanket /ˈblæŋkɪt/ 34
blue /bluː/ 5
boarding pass /bɔːdɪŋ pɑːs/ 36
boat /bəʊt/ 34
book /bʊk/ 4, *5*
bookcase /ˈbʊkkeɪs/ 8
bookshop /ˈbʊkʃɒp/ 14
boring /ˈbɔːrɪŋ/ 15
bottle /ˈbɒt(ə)l/ 26
bottle opener
/ˈbɒt(ə)l ˈəʊpənə(r)/ 34
box /bɒks/ 26
boxing /bɒksɪŋ/ 26-30
boyfriend /ˈbɔɪfrend/ 6, *40*
Brazil /brəˈzɪl/ 1
bread /bred/ 16-20, *17*
break /breɪk/ 33
breakfast /ˈbrekfəst/ 7
bridge /brɪdʒ/ 32
Britain /brɪt(ə)n/ 1
brother /ˈbrʌðə(r)/ 6
brown /braʊn/ 5
Budapest /buːdəpest/ 1, *20*
bunch /bʌntʃ/ 26
bus /bʌs/ 12
bus stop /bʌs stɒp/ 12
business class
/ˈbɪznɪs klɒːs/ 36
butter /ˈbʌtə(r)/ 16-20, *17*
buy /baɪ/ 20

café /ˈkæfeɪ/ 4, *11-15*
cakes /keɪkz/ 26
calculator /ˈkælkjʊˌleɪtə(r)/ 5
camera /ˈkæmrə/ 5
camping /kæmpɪŋ/ 39
camp site /kæmp saɪt/ 28
Canada /ˈkænədə/ 1
car /kɑː(r)/ 12, 33
car park /kɑː(r) pɑːk/ 4, *14, 39*
caravan /ˈkærəˌvæn/ 26-30
carefully /ˈkeəfʊlɪ/ 35
carelessly /ˈkeələslɪ/ 35
carrot /ˈkærət/ 17
carton /ˈkɑːt(ə)n/ 34
cassette /kæˈset/ 4
casual /ˈkæʒʊəl/ 23
catch /kætʃ/ 33
cathedral /kəˈθiːdr(ə)l/ 32
centigrade /ˈsentɪˌgreɪd/ 29
centimetre /ˈsentɪˌmiːtə(r)/ 29
chair /tʃeə(r)/ 4, *8*
change /tʃeɪndʒ/ 24
charges /tʃaːdʒəs/ 39
cheap /tʃiːp/ 12
cheap day return
/tʃiːp daɪ rɪˈtɜːn/ 36
check-in /ˈtʃek ɪn/ 36
cheerful /ˈtʃɪəfʊl/ 16-20
cheese /tʃiːz/ 17
cheesecake /ˈtʃiːzkeɪk/ 25
chemist /ˈkemɪst/ 14
chicken /ˈtʃɪkɪn/ 17
Chinese food
/tʃaɪˈniːz fuːd/ 10
chocolate mousse
/ˈtʃɒkələt muːs/ 25
chocolates /ˈtʃɒkələtz/ 26
cinema /ˈsɪnɪˌmɑː/ 14
classical music
/ˈklæsɪk(ə)l ˈmjuːzɪk/ 10
classroom /ˈklɑːsruːm/ 4
climb /klaɪm/ 32
climbing /klaɪmɪŋ/ 30
cloud /klaʊd/ 37
coach /kəʊtʃ/ 12
coat /kəʊt/ 5
Coca-cola /ˌkəʊkəˌkəʊlə/ 25
coffee /ˈkɒfɪ/ 17
coffee bar /ˈkɒfɪ bɑː(r)/ 40
cold /kəʊld/ 29, *37*
cold (*noun*) /kəʊld/ 28
comb /kəʊm/ 5
comfortable /ˈkʌmftəb(ə)l/ 23
computer /kəmˈpjuːtə(r)/ 4
concerto /kənˈtʃeətəʊ/ 11-15
cook /kʊk/ 13
cook dinner /kʊk ˈdɪnə(r)/ 11
cooker /ˈkʊkə(r)/ 8
cooking /ˈkʊkɪŋ/ 10
cool /kuːl/ 37
cough /kɒf/ 28
cough medicine
/kɒf ˈmedsɪn/ 28
cover /ˈkʌvə(r)/ 31
crash /kræʃ/ 33
crowded /kraʊdɪd/ 12
cup /kʌp/ 34
cupboard /ˈkʌbəd/ 8
curly /ˈkɜːlɪ/ 19
curtains /ˈkɜːt(ə)nz/ 8
cut /kʌt/ 33
cycling /ˈsaɪklɪŋ/ 32

dancing /dɑːnsɪŋ/ 10
dangerous /ˈdeɪndʒərəs/ 30
dark /dɑːk/ 19
daughter /ˈdɔːtə(r)/ 6
December /dɪˈsembə(r)/ 22
decide /dɪˈsaɪd/ 18
departure gate
/dɪˈpɑːtʃə(r) geɪt/ 36
departure lounge
/dɪˈpɑːtʃə(r) laʊndʒ/ 36
detective story
/dɪˈtektɪv ˈstɔːrɪ/ 21
diary /ˈdaɪərɪ/ 5
difficult /ˈdɪfɪkəlt/ 30
dining room /daɪnɪŋ ruːm/ 8
dinner /ˈdɪnə(r)/ 7
director /daɪˈrektə(r)/ 40
disappear /ˌdɪsəˈpɪə(r)/ 21
discover /ˌdɪskʌvə(r))/ 38
dishwasher /ˈdɪʃwɒʃə(r)/ 8
divorced /dɪˈvɔːsd/ 16-20, *21*

dizzy /ˈdɪzɪ/ 28
do /du:/ 13, *20*
do some work/homework /du: sʌm wɜ:k/ həʊmwɜ:k/ 11
do the housework /du: ðə ˈhaʊswɜ:k/ 11
do the washing up /du: ðə ˈwɒʃɪŋ ʌp/ 11
doctor /ˈdɒktə(r)/ 2
double room /ˈdʌb(ə)l ru:m/ 28
downstairs /daʊnˈsteəz/ 8
draw /drɔ:/ 13
dress /dres/ 23
drink coffee /drɪŋk ˈkɒfɪ/ 9
drive /draɪv/ 12
drop /drɒp/ *33*
dry /draɪ/ 29, *37*

eat an apple /i:t æn ˈæp(ə)l/ 9
egg /eg/ 17
eight /eɪt/ 4
eighteen /eɪˈti:n/ 4
eighth /eɪtθ/ 22
eighty /ˈeɪtɪ/ 4
elderly /ˈeldəlɪ/ 19
eleven /ɪˈlev(ə)n/ 4
eleventh /ɪˈlevənθ/ 22
engineer /ˌendʒɪˈnɪə(r)/ 2
England /ˈɪŋglənd/ 20
evening /ˈi:vnɪŋ/ 7
exam /ɪgˈzæm/ 33
exciting /ɪkˈsaɪtɪŋ/ 30
Excuse me /ɪkˈskju:z mi:/ 11-15
expensive /ɪkˈspensɪv/ 12, *30*
eyes /aɪz/ 19

facilities /fæˈsɪlɪtɪs/ 39
fail /feɪl/ 33
faint /feɪnt/ 28
fair /feə(r)/ 19
far /fɑ:(r)/ 12
farmer /ˈfɑ:mə(r)/ 2
fashionable /ˈfæʃnəb(ə)l/ 23, *30*
fast /fɑ:st/ 12, *29*
father /ˈfɑ:ðə(r)/ 6
February /ˈfebrʊərɪ/ 22
ferry /ˈferɪ/ 12
fifteen /fɪfˈti:n/ 4
fifth /fɪfθ/ 22
fifty /ˈfɪftɪ/ 4
find /faɪnd/ 20, *21, 28, 33*
finger /ˈfɪŋgə(r)/ 28, *33*
finish /ˈfɪnɪʃ/ 7
fire /ˈfaɪə(r)/ 8
first /fɜ:st/ 22
first class /fɜ:st klɑ:s/ 36
first name /fɜ:st neɪm/ 3
five /faɪv/ 4

florist /ˈflɒrɪst/ 14
flowers /ˈflaʊə(r)z/ 26
fly /flaɪ/ 20
fog /fɒg/ 37
foggy /ˈfɒgɪ/ 37
foot /fʊt/ 28
football /ˈfʊtbɔ:l/ 10, *30*
foreigner /ˈfɒrɪnə(r)/ 40
forget /fəˈget/ 40
fork /fɔ:k/ 34
forty /ˈfɔ:tɪ/ 4
four /fɔ:(r)/ 4
fourteen /fɔ:ˈti:n/ 4
fourth /fɔ:θ/ 22
France /frɑ:ns/ 20
French fries /frentʃ fraɪz/ 25
fridge /frɪdʒ/ 8
friendly /ˈfrendlɪ/ 4, *16-20*
fruit /fru:t/ 26

garage /ˈgærɑ:dʒ/ 8
garden /ˈgɑ:d(ə)n/ 8
get /get/ 12, *24*
get the bus/train to work/ school/home /get ðə bʌs / treɪn tə wɜ:k / sku:l / həʊm/ 11
get up /get ʌp/ 7
girlfriend /ˈgɜ:lfrend/ 6
glass /glɑ:s/ 27
glasses /glɑ:sɪz/ 5, 19
go /gəʊ/ 20
go by /gəʊ baɪ/ 12
go running /gəʊ ˈrʌnɪŋ/ 9
go shopping /gəʊ ˈʃɒpɪŋ/ 11
go to bed /gəʊ tə bed/ 7
go to the cinema /gəʊ tə ðə ˈsɪnɪˌmɑ:/ 11
go to work/school /gəʊ tə wɜ:k / sku:l/ 7
goal /gəʊl/ 33
going to parties /ˈgəʊɪŋ tə ˈpɑ:tɪz/ 10
going to the cinema /ˈgəʊɪŋ tə ðə ˈsɪnɪˌmɑ:/ 10
going to the theatre /ˈgəʊɪŋ tə ðə ˈθɪətə(r)/ 10
golf /gɒlf/ 30
good /gʊd/ 4
good-looking /gʊd ˈlʊkɪŋ/ 19, *16-20*
grandfather /ˈgrænˌfɑ:ðə(r)/ 6
grandmother /ˈgrænˌmʌðə(r)/ 6
grapes /greɪpz/ 17
green /gri:n/ 5
greengrocer /ˈgri:nˌgrəʊsə(r)/ 14
grey /greɪ/ 5
grow /grəʊ/ 38
guest /gest/ 21

hair /heə(r)/ 19
hamburger /ˈhæmˌbɜ:gə(r)/ 25
hand /hænd/ 28
hang gliding /ˈhæŋˌglaɪdɪŋ/ 30
happy /ˈhæpɪ/ 16
have /hæv/ 20
have a shower /hæv eɪ ˈʃaʊə(r)/ 9
headache /ˈhedeɪk/ 28
heavy /ˈhevɪ/ 27
high /haɪ/ 29
hit record /hɪt ˈrekɔ:d/ 16-20
home /həʊm/ 21
horseriding /ˈhɔ:sraɪdɪŋ/ 30
hot /hɒt/ 29, *37*
hotel /həʊˈtel/ 21, *28*
Hungary /ˈhʌŋgərɪ/ 1, *20*
husband /ˈhʌzbənd/ 6, *21*

ice /aɪs/ 34
ice cream /aɪs kri:m/ 25
ill /ɪl/ 28
in the garden /ɪn ðə ˈgɑ:d(ə)n/ 8
in the mountains /ɪn ðə ˈmaʊntɪns/ 15
indoors /ɪnˈdɔ:z/ 8
industrial /ɪnˈdʌstrɪəl/ 15
interesting /ˈɪntrəstɪŋ/ 4, *15*
international /ˌɪntəˈnæʃən(ə)l/ 4
invent /ɪnˈvent/ 38
invite /ɪnˈvaɪt/ 24
Italy /ˈɪtəli/ 20

jacket /ˈdʒækɪt/ 23
January /ˈdʒænjʊərɪ/ 22
Japan /dʒəˈpæn/ 1
jazz /dʒæz/ 10
jeans /dʒi:nz/ 23, *26*
job /dʒɒb/ 3
judo /ˈdʒu:dəʊ/ 11-15
juice /dʒu:s/ 17
July /dʒu:ˈlaɪ/ 22
June /dʒu:n/ 22

kebab /kɪˈbæb/ 25
keys /ki:z/ 5
kill /kɪl/ 21
kilometre /ˈkɪləˌmi:tə(r)/ 29
kilometres per hour /ˈkɪləˌmi:tə(r)z pɜ:(r) aʊə(r)/ 29
kind /kaɪnd/ 4
kiss /kɪs/ 31
kitchen /ˈkɪtʃɪn/ 8
knife /naɪf/ 34

lamb /læm/ 17
lamp /læmp/ 8
language school /ˈlæŋgwɪdʒ sku:l/ 40

large /lɑːdʒ/ 27
laugh /lɑːf/ 23
lazy /ˈleɪzɪ/ 16
learn /lɜːn/ 18
learn the guitar
/lɜːn ðə gɪˈtɑː(r)/ 9
leather /ˈleðə(r)/ 27
left /left/ 21
leg /leg/ 28
lemon /ˈlemən/ 17
lettuce /ˈletɪs/ 17
library /ˈlaɪbrərɪ/ 4, *14*
lift /lɪft/ 33
light /laɪt/ 27
listen /ˈlɪs(ə)n/ 20
listen to the radio
/ˈlɪs(ə)n tə ðə ˈreidɪəʊ/ 9
live /lɪv laɪv/ 18
lively /laɪvlɪ/ 15
London /ˈlʌnd(ə)n/ 1, 20
long /lɒŋ/ 19, *27*
lose /luːz/ 20, *33*
loudly /laʊdlɪ/ 37
love /lʌv/ 40
lovely /ˈlʌvlɪ/ 15
low /ləʊ/ 29
lunch /lʌntʃ/ 7

magazines /ˌmægəˈziːnz/ 10
make /meɪk/ 38
make friends /meɪk frendz/ 20
March /mɑːtʃ/ 22
market /ˈmɑːkɪt/ 14
marriage /ˈmærɪdʒ/ 16-20, *21*
married /ˈmærɪd/ 3
marry /ˈmærɪ/ 18
matches /mætʃɪz/ 34
May /meɪ/ 22
mayonnaise /ˌmeɪəˈneɪz/ 25
meal /miːl/ 16-20
medium-height
/ˈmiːdɪəm haɪt/ 19
metal /ˈmet(ə)l/ 27
metre /ˈmiːtə(r)/ 29
Mexico /ˈmeksɪkəʊ/ 1
midday /ˈmɪddeɪ/ 7
middle-aged
/ˈmɪd(ə)l eɪdʒd/ 19
millimetre /ˈmɪlɪˌmiːtə(r)/ 29
miss /mɪs/ 33, *40*
mobile phone
/ˈməʊbaɪl fəʊn/ 5
modern /ˈmɒd(ə)n/ 15
morning /ˈmɔːnɪŋ/ 7
mother /ˈmʌðə(r)/ 6
motor racing
/ˈməʊtə(r) reɪsɪŋ/ 30
motorbike /ˈməʊtə(r) baɪk/ 12
moustache /məˈstɑːʃ/ 19
mystery /ˈmɪstərɪ/ 21

naughty /ˈnɔːtɪ/ 16
near /nɪə(r)/ 12
newsagent /ˈnjuːzˌeɪdʒ(ə)nt/ 14
night /naɪt/ 7
nine /naɪn/ 4
nineteen /naɪntiːn/ 4
ninety one /ˈnaɪntɪ wʌn/ 4
ninth /naɪnθ/ 22
north east /nɔːθ iːst/ 15
north west /nɔːθ west/ 15
notebook /ˈnəʊtbʊk/ 5
November /nəˈvembə(r)/ 22
nylon /ˈnaɪlɒn/ 27

October /ɒkˈtəʊbə(r)/ 22
oil /ɔɪl/ 17
old /əʊld/ 15, *19, 29*
on an island
/ɒn æn ˈaɪlənd/ 15
on the coast /ɒn ðə kəʊst/ 15
on the river /ɒn ðə ˈrɪvə(r)/ 15
one hundred
/wʌn ˈhʌndrəd/ 4
onion /ˈʌnjən/ 17
orange /ˈɒrɪndʒ/ 17
outdoors /aʊtˈdɔːz/ 8

packet /ˈpækɪt/ 26
paint /peint/ 38
pair /peə(r)/ 26
paper /ˈpeɪpə(r)/ 27
Pardon /ˈpɑːd(ə)n/ 11-15
Paris /ˈpærɪs/ 20
park /pɑːk/ 32
pass /pɑːs/ 33
passport control
/ˈpɑːspɔːt kənˈtrəʊl/ 36
pasta /ˈpæstə/ 25
pay /peɪ/ 33
peach /piːtʃ/ 17
pen /pen/ 5
pencil /ˈpensɪl/ 5
perfume /ˈpɜːfjuːm/ 26
personal stereo
/ˈpɜːsən(ə)l ˈsterɪəʊ/ 5
phone box /fəʊn bɒks/ 14
phone number
/fəʊn ˈnʌmbə(r)/ 3
pink /pɪŋk/ 5
pizza /ˈpiːtsə/ 25
plane /pleɪn/ 12, *36*
plastic /ˈplæstɪk/ 27
plate /pleɪt/ 33
platform /ˈplætfɔːm/ 36
play /pleɪ/ 13, *18*
play football /pleɪ ˈfʊtbɔːl/ 9
player /ˈpleɪə(r)/ 4
Please /pliːz/ 11-15
point at /pɔɪnt æt/ 31
police officer
/pəˈliːs ˈɒfɪsə(r)/ 2

polite /pəˈlaɪt/ 16-20
politely /pəˈlaɪtlɪ/ 35
popular /ˈpɒpjʊlə(r)/ 4, *30*
post office /pəʊst ˈɒfɪs/ 14
potato /pəˈteɪtəʊ/ 17
pretty /ˈprɪtɪ/ 19
pub /pʌb/ 14
purple /ˈpɜːp(ə)l/ 5

quickly /kwɪklɪ/ 35
quietly /ˈkwaɪətlɪ/ 35

railway station
/ˈreɪlweɪ ˈsteɪʃ(ə)n/ 14
rain /reɪn/ 37
rainy /reɪnɪ/ 37
read /riːd/ 20
read a newspaper
/riːd ə ˈnjuːsˌpeɪpə(r)/ 9
receive /rɪˈsiːv/ 18
reception desk
/rɪˈsepʃ(ə)n desk/ 4
receptionist /rɪˈsepʃənɪst/ 2
rectangular
/rekˈtæŋgʊlə(r)/ 27
red /red/ 5
restaurant /ˈrestəˌrɒnt/ 14, *16-20*
return /rɪˈtɜːn/ 36
rice /raɪs/ 17
ride /raɪd/ 12, *13*
ring /rɪŋ/ 5
risotto /rɪˈzɒtəʊ/ 25
rock music /rɒk ˈmjuːzɪk/ 10
room service /ruːm ˈsɜːvɪs/ 28
round /raʊnd/ 27
rubbish /ˈrʌbɪʃ/ 34
rudely /ruːdlɪ/ 35

salad /ˈsæləd/ 25
samba /ˈsæmbə/ 11-15
sandwich /ˈsænwɪdʒ/ 25
sauna /ˈsɔːnə/ 11-15
save /seɪv/ 24
score /skɔː(r)/ 33
sea /siː/ 40
second /ˈsekənd/ 22
secretary /ˈsekrətrɪ/ 2
see friends /siː frendz/ 11
self–study room /self ˈstʌdɪ/ 4
September /sepˈtembə(r)/ 22
serious /ˈsɪərɪəs/ 16
seven /ˈsev(ə)n/ 4
seventeen /ˌsevənˈtiːn/ 4
seventh /ˈsev(ə)nθ/ 22
seventy /ˈsevəntɪ/ 4
shake hands /ʃeɪk hænds/ 31
shirt /ʃɜːt/ 23
shoes /ʃuːz/ 23
short /ʃɔːt/ 19, *27*
shorts /ʃɔːtz/ 23

shower /ˈʃauə(r)/ 8
shy /ʃaɪ/ 16
sick /sɪk/ 28
siesta /srˈesta/ 11-15
single /ˈsɪŋg(ə)l/ 36
single room
/ˈsɪŋg(ə)l ru:m/ 28
sister /ˈsɪstə(r)/ 6
sit down /sɪt daʊn/ 23
sitting room /sɪtɪŋ ru:m/ 8
six /sɪks/ 4
sixteen /ˌsɪksˈti:n/ 4
sixth /sɪksθ/ 22
sixty /ˈsɪkstɪ/ 4
ski /ski:/ 13
skiing /ski:ɪŋ/ 30
skirt /skɜ:t/ 23
slim /slɪm/ 19
slow /sləʊ/ 12, *29*
slowly /sləʊlɪ/ 35
small /smɔ:l/ 15, *19, 27, 29*
smart /smɑ:t/ 23
smile /smaɪl/ 23
snow /snəʊ/ 37
soap /səʊp/ 26
socks /sɒks/ 23
sofa /ˈsəʊfə/ 8
son /sʌn/ 6
sore throat /sɔ:(r) θrəʊt/ 28
Sorry /ˈsɒrɪ/ 11-15
south east /saʊθ i:st/ 15
south west /saʊθ west/ 15
spaghetti /spəˈgetɪ/ 11-15
speak /spi:k/ 13
spend /spend/ 24
square /skweə(r)/ 27, *29*
stand /stænd/ 23
start /stɑ:t/ 7, *18*
station /ˈsteɪʃ(ə)n/ 12, *14*
stay /steɪ/ 20
steak /steɪk/ 25
steal /sti:l/ 33
step /step/ 40
stomach ache
/ˈstʌmək eɪk/ 28
straight /streɪt/ 19
strange /streɪndʒ/ 40
strawberry /ˈstrɔ:bərɪ/ 25
stubborn /ˈstʌbən/ 16
student /ˈstju:d(ə)nt/ 2
study /ˈstʌdɪ/ 8
successful /səkˈsesfʊl/ 21
suit /su:t/ 23
sun /sʌn/ 37
sunny /sʌnɪ/ 37
supermarket /ˈsu:pəˌmɑ:kɪt/ 14
surname /ˈsɜ:neɪm/ 3
sweater /ˈswetə(r)/ 23
swim /swɪm/ 13
swimming /swɪmɪŋ/ 10, *30*
swimsuit /ˈswɪmsu:t/ 23
Switzerland /swɪtsəlænd/ 20

table /ˈteɪb(ə)l/ 4, *8*
take /teɪk/ 12, *20, 24*
take off /teɪk ɒf/ 31
tall /tɔ:l/ 19
taxi driver /ˈtæksɪ ˈdraɪvə(r)/ 2
taxi rank /ˈtæksɪ ræŋk/ 14
tea /ti:/ 17
teacher /ˈti:tʃə(r)/ 2
television /ˈtelɪˌvɪʒ(ə)n/ 8
tell /tel/ 21
temperature /ˈtemprɪtʃə(r)/ 28
tennis /ˈtenɪs/ 10, *30*
tent /tent/ 28
tenth /tenθ/ 22
Thailand /taɪlænd/ 1
Thank you /θæŋk ju:/ 11-15
third /θɜ:d/ 22
thirteen /θɜ:ˈti:n/ 4
thirty /ˈθɜ:tɪ/ 4
three /θri:/ 4
ticket office /ˈtɪkɪt ˈɒfɪs/ 36
tie /taɪ/ 23
tights /taɪts/ 23
tired /ˈtaɪəd/ 28
tiring /ˈtaɪərɪŋ/ 30
toe /təʊ/ 28
toilet /ˈtɔɪlɪt/ 8
Tokyo /ˈtəʊkɪəʊ/ 1
tomato /təˈmɑ:təʊ/ 17
tourist class /ˈtʊərɪst klɑ:s/ 36
train /treɪn/ 12, *33, 36*
trainers /ˈtreɪnə(r)s/ 23
tram /ˈtræm/ 12
travel /ˈtræv(ə)l/ 39
trousers /ˈtraʊzəz/ 23
twelth /twelfθ/ 22
type /taɪp/ 13
T-shirt /ti: ʃɜ:t/ 23

ugly /ˈʌglɪ/ 15
uncle /ˈʌŋk(ə)l/ 6
underground
/ˌʌndəˈgraʊnd/ 12
unhappy /ʌnˈhæpɪ/ 21
upstairs /ʌpˈsteəz/ 8
use /ju:z/ 13

Venice /ˈvenɪs/ 20
vet /vet/ 2
Vienna /vɪːənə/ 20
view /vju:/ 32
visit /ˈvɪzɪt/ 20

waiter /ˈweɪtə(r)/ 2, *21*
wake-up call
/weɪk ʌp kɔ:l/ 28
wallet /ˈwɒlɪt/ 5, *33*
warm /wɔ:m/ 23, *37*
washbasin /wɒʃ ˈbeɪs(ə)n/ 8
watch /wɒtʃ/ 5, *20*
watch a tennis match
/wɒtʃ ə tenɪs mætʃ/ 9
watching sport
/wɒtʃɪŋ spɔ:t/ 10
water /ˈwɔ:tə(r)/ 17
wear /weə(r)/ 23
weekend /wi:kˈend/ 7
well /wel/ 35
well-behaved
/wel bɪˈheɪvd/ 16-20
well-built /wel bɪlt/ 19
wet /wet/ 29, *37*
white /waɪt/ 5
wife /waɪf/ 6
wind /wɪnd/ 37
windsurfing /ˈwɪndˌsɜ:fɪŋ/ 30
windy /wɪndɪ/ 37
wine /waɪn/ 17, *25, 26*
wood /wʊd/ 27
work /wɜ:k/ 7, *18*
write /raɪt/ 13, *20, 38*
writer /ˈraɪtə(r)/ 21

yellow /ˈjeləʊ/ 5
young /jʌŋ/ 19, *29*
youth hostel /ju:θ ˈhɒst(ə)l/ 39

Zurich /ˈzju:rɪk/ 20

Progress Test 1 Lessons 1 – 10

SECTION 1: VOCABULARY (30 marks)

1a Underline the odd-one-out and leave a group of three related words. (10 marks)

b Add one other word to the groups of words. (10 marks)

Example: her my our they *their*

1 French Canada Hungary Thailand ____________
2 journalist secretary taxi waiter ____________
3 eight number seventeen thirty ____________
4 friendly interesting kind person ____________
5 beautiful orange purple yellow ____________
6 aunt daughter family brother ____________
7 bedroom home kitchen study ____________
8 bed books chair cupboard ____________
9 computer television upstairs video ____________
10 babies men people woman ____________

2 Complete these sentences with ten different verbs. (10 marks)

Example: I *read* a newspaper every morning.

1 I ____________ with my husband in north London.
2 I ____________ in a school. I'm a teacher.
3 I ____________ up at seven in the morning.
4 I ____________ to school. Our house is near the school.
5 I ____________ work at four o'clock in the afternoon.
6 My husband and I ____________ dinner at eight o'clock.
7 Then I ____________ television.
8 I ____________ to bed at eleven o'clock.
9 I ____________ toast for breakfast.
10 I ____________ tennis with my husband.

Progress Test 1 Lessons 1 – 10

SECTION 2: GRAMMAR (30 marks)

3a Choose ten of these words to complete the first ten spaces in the passage. (10 marks)

Example: a) does b) has c) <u>is</u>

1 a) Her b) His c) Their

2 a) from b) in c) on

3 a) a lot b) at all c) not at all

4 a) a b) an c) any

5 a) people b) peoples c) person

6 a) a b) any c) some

7 a) a b) any c) some

8 a) watch b) watches c) watching

9 a) doesn't b) hasn't c) isn't

10 a) Their b) They c) They're

b Complete the last ten spaces with ten of your own words. (10 marks)

I've got a sister and a brother. My sister *is* twenty. (1) ___________ name's Barbara. She's at university (2) ___________ Oxford. Barbara likes Oxford (3) ___________. It is (4) ___________ lovely city and the (5) ___________ are very nice. Barbara's got a small room in a college. There's a desk, (6) ___________ chairs and a cupboard in her room. There's a mirror but there aren't (7) ___________ plants. Barbara hasn't got a television – she doesn't like (8) ___________ it. Barbara (9) ___________ married but she's got a boyfriend. (10) ___________ very happy.

My brother's name (11) ___________ Bill. He's (12) ___________ artist. Bill stays (13) ___________ home every day. He works in (14) ___________ study. Bill and his wife (15) ___________ got two daughters. (16) ___________ names are Lisa and Jane. They (17) ___________ five years old. In Britain children start school (18) ___________ five. Lisa and Jane are at school. Lisa doesn't like school (19) ___________ much but her sister likes (20) ___________ a lot.

4 Write your own questions for these answers. (10 marks)

Example: My name's Pat.
What's your name?

1 Fine, thanks.

____

2 I'm from London.

____

3 It's 25, Hill Avenue.

____

4 It's 0171 647 8902.

____

5 I'm a teacher.

____

6 I'm twenty-five.

____

7 Yes, I am. My husband's name is Greg.

____

8 Yes, I've got three children.

____

9 It's blue.

____

10 It's half past one.

____

Progress Test 1 Lessons 1 – 10

SECTION 3: READING (20 marks)

5 Read the passage *How do you relax?* about Bill and Kate and their daughter, Sally.
Complete the sentences. (10 marks)

Example: *Bill* likes watching football.

1 ________ likes doing some sport.

2 ________ like going to the cinema.

3 ________ likes pop music.

4 ________ like rock music.

5 ________ likes seeing friends.

6 Answer the questions. (10 marks)

Example: Who does Bill live with?
He lives with his wife.

1 Where does Bill's wife watch TV?

____

2 When does Sally go to a club?

____

3 Why does Kate get some Chinese food?

____

4 What does Bill do at the weekend?

____

5 Who does Kate go to the cinema with?

____

HOW DO YOU RELAX?

I like football. I play once a week, at the weekend. I like watching it on television too. My wife, Kate, and I have got two televisions. She doesn't like watching football at all. I watch football downstairs and she watches films in our bedroom. I like rock music too – and so does Kate.
Bill

My husband and I have got one daughter. We don't usually go out in the evening. Sometimes we get a video and some Chinese food – I don't like cooking. We both like films. Once a month my sister stays with us. She stays at home with our teenage daughter, Sally, while we go to the cinema. I like going to the cinema.
Kate

I like listening to music. I like pop music and jazz but I don't like rock. At the weekend I go to a jazz club with my friends. Courtney Pine is my favourite musician. I like going to the cinema and theatre too.
Sally

Progress Test 1 Lessons 1 – 10

SECTION 4: WRITING (20 marks)

7 Write about your home. Write about the rooms and furniture. Write 8 – 10 sentences. (20 marks)

Progress Test 2 Lessons 11 – 20

SECTION 1: VOCABULARY (30 marks)

1a Underline the odd-one-out and leave a group of three related words. (10 marks)

b Add one other word to the groups of words. (10 marks)

Example: her my our <u>they</u> *their*

1 after always never usually ___________

2 bus plane train transport ___________

3 cheap crowded fast ferry ___________

4 chemist greengrocer newsagent station ___________

5 coast north south west ___________

6 beautiful boring small town ___________

7 favourite quiet shy stubborn ___________

8 coffee juice oil water ___________

9 curly dark hair long ___________

10 fly knit run swim ___________

2 Complete these sentences with ten different verbs. (10 marks)

Example: You can *get* a taxi from a taxi rank.

1 I can ___________ a car.

2 She can ___________ crosswords.

3 I can ___________ a foreign language.

4 You can ___________ some flowers at the florist.

5 I can ___________ a meal for six people.

6 You can ___________ a film at the cinema.

7 I can ___________ a horse.

8 You can ___________ a book from the library.

9 I can ___________ a musical instrument.

10 He can ___________ a computer.

Progress Test 2 Lessons 11 – 20

SECTION 2: GRAMMAR (30 marks)

3a Choose ten of these words to complete the first ten spaces in the conversation. (10 marks)

Example: a) from b) of c) to

1 a) Their b) They c) They're

2 a) from b) in c) of

3 a) stay b) stayed c) staying

4 a) along b) opposite c) next

5 a) has b) is c) was

6 a) be b) go c) visit

7 a) visited b) was c) went

8 a) Am b) Are c) Is

9 a) boring b) interesting c) wonderful

10 a) by b) in c) on

b Complete the last ten spaces with ten of your own words. (10 marks)

JAMIE: Look! I've got a postcard *from* Sue and Sam. (1) ___________ on holiday in Albi for a week.

CHRIS: Albi? Where's Albi?

JAMIE: It's in the south-west (2) ___________ France. Sue says they're (3) ___________ in a hotel in the main square. It's (4) ___________ the cathedral. Albi's a lovely town. I (5) ___________ there last year.

CHRIS: Did you (6) ___________ to the cathedral?

JAMIE: Yes, I did. And I (7) ___________ the Toulouse Lautrec museum.

CHRIS: (8) ___________ Sue and Sam enjoying themselves?

JAMIE: Yes, they're having a (9) ___________ time. They went (10) ___________ plane to Toulouse and then they hired a car.

CHRIS: What (11) ___________ we need for dinner this evening?

JAMIE: We've (12) ___________ some chicken in the fridge. Let's get (13) ___________ vegetables.

CHRIS: (14) ___________ about potatoes and carrots?

JAMIE: Yes, OK. And we haven't got (15) ___________ onions. Let's get some onions, too.

CHRIS: Let's (16) ___________ to the new supermarket.

JAMIE: Where is it?

CHRIS: It's (17) ___________ South Road. It's (18) ___________ the right. It's next to the department store.

JAMIE: Oh, yes. (19) ___________ you got any money?

CHRIS: Yes, I went to the bank this morning and took (20) ___________ some money.

4 Write your own questions for these answers. (10 marks)

Example: My name's Pat.
What's your name?

1 I go by bus.

____

2 It takes about fifteen minutes.

____

3 Eighty pence. It's quite cheap.

____

4 It's about six kilometres.

____

5 He's one metre eighty.

____

6 I'm twenty-five.

____

7 He's very nice.

____

8 He's tall with curly hair and glasses.

____

9 I get up at half past seven.

____

10 Knit? No, I can't.

____

Progress Test 2 Lessons 11 – 20

SECTION 3: READING (20 marks)

5 Read the passage *A Day in the Life.*
Complete the sentences. (10 marks)

Example: She *lives* in a flat.

1 She has __________ before breakfast.

2 She drinks __________ for breakfast.

3 She has __________ after her driving lesson.

4 She has __________ after she gets home from the theatre.

5 She goes __________ at about three o'clock.

6 Are these sentences true (T) or false (F) or doesn't the passage say (DS)? (10 marks)

Example: Clare hasn't got any brothers. [DS]

1 She was born in London. ☐

2 She gets up late. ☐

3 She goes to the gym in the afternoon. ☐

4 'On the road' is at a theatre in London. ☐

5 She drives to the theatre. ☐

A DAY IN THE LIFE

Clare Smith was born in 1976, the third daughter of the actor, Charles Smith. As a child, Clare worked in television and the theatre. In 1992 Clare left school to follow her acting career. She is currently appearing in the musical, 'On the road'.

I live in London. I've got a small flat in north London. I don't get up very early – about eleven o'clock. This is because I work in the evening and go to bed late. When I get up I have a shower. Then I have breakfast – a cup of coffee, brown toast and some fruit. After breakfast I go to the gym for about an hour. Then I have a driving lesson. After that I have lunch, usually salad or some pasta. I don't have a big lunch. I relax after lunch. I watch a good video or I sometimes go shopping. Then I go to the theatre by underground. I'm in the musical 'On the road' at the moment. I have dinner when I get home from the theatre. The evening show is at half past seven and I get home at about eleven o'clock. I have dinner at about half past eleven. I go to bed at about two o'clock. What time do I go to sleep? Well, at about three o'clock.

Progress Test 2 Lessons 11 – 20

SECTION 4: WRITING (20 marks)

7 Write about two of your friends. What are they like? What do they look like? What do they do? Write 8 – 10 sentences. (20 marks)

Progress Test 3 Lessons 21 – 30

SECTION 1: VOCABULARY (30 marks)

1a Underline the odd-one-out and leave a group of three related words. (10 marks)

b Add one other word to the groups of words. (10 marks)

Example: her my our <u>they</u> *their*

1 eight first ninth third ___________
2 August March Monday July ___________
3 casual socks sweater tie ___________
4 coffee kebab pizza steak ___________
5 black light round short ___________
6 bag glass paper plastic ___________
7 his mine whose yours ___________
8 cough headache medicine temperature ___________
9 best biggest fast worst ___________
10 basketball cycling knitting skiing ___________

2 Complete these sentences with ten different verbs. (10 marks)

Example: I *work* in an office.

1 I ___________ stories in my free time.
2 I ___________ to the dentist twice a year.
3 I ___________ comfortable clothes.
4 I ___________ a lot of exercise.
5 I ___________ eight hours a day at work.
6 I ___________ hands when I meet people.
7 I ___________ gifts for my friends on their birthdays.
8 I ___________ sick.
9 I ___________ plenty of water.
10 I ___________ hang gliding is dangerous.

Progress Test 3 Lessons 21 – 30

SECTION 2: GRAMMAR (30 marks)

3a Choose ten of these words to complete the first ten spaces in the conversation. (10 marks)

Example: a) Are b) Do c) Would

1 a) I'm not b) I don't c) I wouldn't
2 a) a b) any c) some
3 a) they b) we c) you
4 a) You b) You're c) You've
5 a) me b) mine c) my
6 a) did b) do c) will
7 a) in b) next c) on
8 a) a b) on c) the
9 a) from b) of c) with
10 a) because b) but c) so

b Complete the last ten spaces with ten of your own words. (10 marks)

CHRIS: *Would* you like a cup of coffee?
JANE: No, thank you. (1) ___________ like coffee. Can I have (2) ___________ water?
CHRIS: Yes, here (3) ___________ are. I need a new raincoat.
JANE: (4) ___________ got a raincoat.
CHRIS: No, I haven't. I lost (5) ___________ in August.
JANE: Where (6) ___________ you lose it?
CHRIS: I lost it on holiday. It was in my bag and I left it (7) ___________ a chair at the airport.
JANE: What (8) ___________ pity! It was a nice coat.
CHRIS: Yes, it was. It was very light.
JANE: What was it made (9) ___________ ?
CHRIS: Nylon. I bought it (10)___________ it was light.

ASSISTANT: Good afternoon. (11) ___________ I help you?
CHRIS: Yes, I'm looking (12) ___________ a raincoat.
ASSISTANT: We've got some raincoats (13) ___________ there.
CHRIS: This brown one is nice. Can I try it (14) ___________ ?
ASSISTANT: Yes, go (15) ___________.
CHRIS: No, it doesn't fit (16) ___________. Have you got one in a bigger size?
ASSISTANT: No, I'm afraid (17) ___________. That's the (18) ___________ one in that colour.
CHRIS: Have you got it (19) ___________ green?
ASSISTANT: Not in a larger size. What about the black one?
CHRIS: No, black doesn't suit me. I'll leave (20) ___________. Thank you.

4 Write these sentences in the negative. (10 marks)

Example: I can speak French.
I can't speak French.

1 She's got an aspirin.
____

2 He found his coat.
____

3 They had a happy marriage.
____

4 We like horseriding.
____

5 I'm older than John.
____

6 He's wearing brown shoes.
____

7 You should sit down.
____

8 She does the housework.
____

9 It was wet.
____

10 We're going away this year.
____

Progress Test 3 Lessons 21 – 30

SECTION 3: READING (20 marks)

5 Read the passage *The Olympic Games.* Are these sentences true (T) or false (F) or doesn't the passage say (DS)? (10 marks)

Example: The ancient games took place in different cities. F

1 Women took part in the first modern Olympic games. ☐

2 The ancient games included ball games. ☐

3 There were no games for about 1,500 years. ☐

4 Norma Enriqueta Basilio de Sotela won a medal. ☐

5 Mark Spitz swam in more than one Olympic games. ☐

6 Answer the questions. (10 marks)

Example: Where were the first Olympic games?
They were in Olympia, in Greece.

1 What happened for the first time in 1968?

____

2 How many teams were at the Atlanta games?

____

3 Why are there five rings in the Olympic symbol?

____

4 What happened about a hundred years ago?

____

5 In which year did beach volleyball first become an Olympic sport?

____

THE OLYMPIC GAMES

* The Olympic games began in Greece more than 2,000 years ago. Young men from different cities took part in running, horse-riding, boxing and riding. No women took part in or watched the games.
* The last ancient games took place in the year 394 and for many hundreds of years there were no games. Then, in 1896 the first modern Olympic games took place, again in Greece. Most countries send a team of athletes to the games. In the 1996 games, 197 countries sent teams to Atlanta. Beach volleyball became an Olympic sport for the first time.
* At the start of the Olympic games, an athlete runs into the Olympic stadium with a torch and lights an Olympic flame. The flame comes all the way from Olympia. The lighting of the torch is part of an enormous opening ceremony. In 1968 Mexican hurdler, Norma Enriqueta Basilio de Sotela, became the first woman to light the Olympic flame.
* The Olympic symbol is five coloured rings. These rings represent the five continents in the world.
* Winners receive gold, silver and bronze medals. Sometimes competitors win more than one medal. In 1972, Mark Spitz, an American swimmer, won seven gold medals. He also broke four world records.

Progress Test 3 Lessons 21 – 30

SECTION 4: WRITING (20 marks)

7 Write a short autobiography. Write about some special events in your life. Write 8 – 10 sentences. (20 marks)

Progress Test 4 Lessons 31 – 40

SECTION 1: VOCABULARY (30 marks)

1a Write two words from the box in each category below. (10 marks)

aunt chemist chair China cough cupboard cycling December ferry florist golf headache Hungary journalist March plane salad sister sweater tie waiter water

Example:

clothes	*sweater*	*tie*
complaints		
countries		
family		
food and drink		
furniture		
jobs		
means of transport		
months		
shops		
sports		

b Add one other word to the categories. (10 marks)

clothes	
complaints	
countries	
family	
food and drink	
furniture	
jobs	
means of transport	
months	
shops	
sports	

2 Complete these sentences with ten different verbs. (10 marks)

Example: I *have* lunch at home.

1 I __________ photographs with my camera.
2 I often __________ in hotels when I go on holiday.
3 I __________ to music on my radio.
4 I always __________ my bills every month.
5 I __________ a knife to cut meat.
6 I usually __________ a list before I go shopping.
7 I __________ clearly so everyone can hear me.
8 I don't __________ exams very easily.
9 I __________ hard at school.
10 I sometimes __________ a goal when I play football.

Progress Test 4 Lessons 31 – 40

SECTION 2: GRAMMAR (30 marks)

3a Choose ten of these words to complete the first ten spaces in the conversation. (10 marks)

Example: a) phones b) phone's c) phones'

1 a) I b) I'll c) I'm

2 a) How b) Where c) Who

3 a) at b) in c) on

4 a) not b) so c) very

5 a) ever b) never c) yet

6 a) has been b) is c) was

7 a) ago b) last c) past

8 a) go b) have gone c) went

9 a) are you staying b) do you stay c) have you stayed

10 a) in b) on c) this

b Complete the last ten spaces with ten of your own words. (10 marks)

JAMIE: Chris! The *phone's* ringing.

CHRIS: OK, (1) ___________ answer it. ... Hello, 379462.

JANE: Hi, Chris! It's Jane.

CHRIS: Jane! (2) ___________ are you?

JANE: I'm fine. I'm in the USA. I'm in New York with Beth (3) ___________ the moment. We've been here two days but we've done (4) ___________ much already. Have you (5) ___________ been to the States?

CHRIS: Yes, I have.

JANE: When (6) ___________ that?

CHRIS: Three years (7) ___________. I (8) ___________ skiing in Colorado. It was great. But I haven't been to New York. Where (9) ___________?

JANE: In a hotel on Fifth Avenue. It's near Central Park. We went there (10) ___________ morning. Central Park can be dangerous but we (11) ___________ careful.

CHRIS: Have you been to the Statue of Liberty? I've (12) ___________ finished reading an article about it. It (13) ___________ opened in 1885. (14) ___________ you know that?

JANE: No, I didn't. We'll probably go there (15) ___________ Monday morning. There's a ferry from Manhattan (16) ___________ half hour.

CHRIS: How's Beth? Is she (17) ___________ a good time?

JANE: She's (18) ___________ a cold. I think she (19) ___________ take some aspirin, but she won't. She says it'll (20) ___________ better tomorrow.

4 Underline the correct expression in the sentences below. (10 marks)

Example: I'd like *dancing/to dance*.

1 *I/I've* moved to London when I was eighteen.

2 *I/I'd* like a ticket to Paris.

3 The Taj Mahal *built/was built* by Shah Jehan.

4 He said he *had/has* read the book.

5 It *takes/is taking* ten minutes by car.

6 He *doesn't speak/isn't speaking* English very well.

7 My birthday *has been/was* in June.

8 I *didn't finish/haven't finished* my homework yet.

9 My back *hurt/hurts* when I sat down.

10 *I'm wearing/I wear* trainers today.

Progress Test 4 Lessons 31 – 40

SECTION 3: READING (20 marks)

5 Read the passage *Rice*.
Tick the correct answers. (10 marks)

Example: People in China

a) are enormous.
b) eat a lot of rice. ✓
c) greet each other with rice.
d) need a lot of water.

1 Most of the world's rice is grown
a) in Australia.
b) in Asia.
c) on hills.
d) in Japan.

2 Asian countries
a) always produce enough rice.
b) don't have any flat land.
c) produce more potatoes than rice.
d) sometimes have to import rice.

3 Rice is planted
a) by machine in Asia.
b) by hand and picked by machine.
c) by hand in the United States.
d) by hand in most Asian countries.

4 North Americans
a) don't like potatoes.
b) eat more bread than rice.
c) don't eat rice.
d) prefer rice from China.

5 Rice was grown in Egypt
a) before it was grown in China.
b) after it was grown in Italy.
c) before it was grown in Europe.
d) after it was grown in the USA.

RICE

Rice is one of the most important foods in the world. This is because people who live in China, Japan and India and other parts of Asia, live mainly on rice. In China the word 'rice' is used in one of their greetings. People say 'Have you eaten your rice today?' This is because rice is so important to the Chinese.

Although most of the world's rice is produced in Asia, sometimes it has to be imported. This happens when rice doesn't grow properly. If there is no rice harvest, people in Asia may die.

Rice was first grown in China about five thousand years ago and it was then introduced into Egypt.

Rice was first grown in Europe, in Italy, about six hundred years ago. It was not grown until three hundred years ago in the United States. North Americans eat some rice but they prefer bread and potatoes.

Some rice is grown on hills but most rice is grown on flat land near lakes and rivers because rice needs a lot of water. In the rice-growing parts of the United States, rice is planted and picked by machine but in most Asian countries everything is done by hand.

6 Are these sentences true (T) or false (F) or doesn't the passage say (DS)? (10 marks)

Example: All Asian people eat 400-800 kilograms of rice a year. DS

1 North Americans eat more rice than Europeans. ☐
2 Rice is grown near lakes and rivers because it needs a lot of water. ☐
3 Rice was probably grown in Egypt five thousand years ago. ☐
4 North Americans have just started eating rice. ☐
5 Machines are never used to plant or pick rice in Asia. ☐

Progress Test 4 Lessons 31 – 40

SECTION 4: WRITING (20 marks)

7 Write about one day last week. Write about something special or unusual you did. Write 8 – 10 sentences. (20 marks)

Answers Progress Test 1 Lessons 1 – 10

SECTION 1: VOCABULARY [30 marks]

1a (10 marks: 1 mark for each correct answer.)

1 French
2 taxi
3 number
4 person
5 beautiful
6 family
7 home
8 books
9 upstairs
10 woman

b (10 marks: 1 mark for each appropriate answer.)

1 a country, eg *Japan, France*
2 a job, eg *farmer, actor*
3 a number, eg *four, twenty*
4 an adjective, eg *good, popular*
5 a colour, eg *grey, pink*
6 a family member, eg *grandmother, sister*
7 a room or place in the home, eg *dining room, bedroom*
8 an item of furniture, eg *bookcase, sofa*
9 an item of household equipment, eg *stereo system, telephone*
10 a plural noun, eg *children, curtains*

2 (10 marks: 1 mark for each appropriate answer.)
Possible answers

1 live
2 work
3 get
4 walk
5 finish
6 have
7 watch
8 go
9 have/eat
10 play

SECTION 2: GRAMMAR [30 marks]

3a (10 marks: 1 mark for each correct answer.)

1 a) her
2 b) in
3 a) a lot
4 a) a
5 a) people
6 c) some
7 b) any
8 c) watching
9 c) isn't
10 c) they're

b (10 marks: 1 mark for each appropriate answer.)
Possible answers

11 is
12 an
13 at
14 his/the
15 have
16 Their
17 are
18 at
19 very
20 it

4 (10 marks: 1 mark for each correct question.)
Possible answers

1 How are you?
2 Where are you from?
3 What's your address?
4 What's your telephone number?
5 What's your job?/What do you do?
6 How old are you?
7 Are you married?
8 Have you got any children?
9 What colour is it?/What's your favourite colour?
10 What time is it?/What's the time?

SECTION 3: READING [20 marks]

5 (10 marks: 2 marks for each correct answer.)

1 Bill 2 Sally and Kate 3 Sally 4 Bill and Kate 5 Sally

6 (10 marks: 2 marks for each correct answer.)

1 (She watches television) in their bedroom.
2 (She goes to a club) at the weekend.
3 (She gets some Chinese food because) she doesn't like cooking.
4 He plays football (at the weekend).
5 (She goes to the cinema) with her husband, Bill.

SECTION 4: WRITING [20 marks]

7 (20 marks)

Tell students what you will take into consideration when marking their written work. Criteria should include:

* efficient communication of meaning (7 marks)
* grammatical accuracy (7 marks)
* coherence in the ordering of the information or ideas (3 marks)
* layout, capitalisation and punctuation (3 marks)

It is probably better not to use a rigid marking system with the written part of the test. If, for example, you always deduct a mark for a grammatical mistake, you may find that you are over-penalising students who write a lot or who take risks. Deduct marks if students haven't written the minimum number of sentences stated in the test.

Answers Progress Test 2 Lessons 11 – 20

SECTION 1: VOCABULARY [30 marks]

1a (10 marks: 1 mark for each correct answer.)

1 after
2 transport
3 ferry
4 station
5 coast
6 town
7 favourite
8 oil
9 hair
10 knit

b (10 marks: 1 mark for each appropriate answer.)

1 an adverb of frequency, eg *often, sometimes*
2 a means of transport, eg *bicycle, car*
3 an adjective, eg *slow, expensive*
4 a shop, eg *baker, florist*
5 a point on the compass, eg *east*
6 an adjective, eg *lovely, modern*
7 an adjective of character, eg *friendly, lazy*
8 a drink, eg *coke, tea*
9 an adjective for describing hair, eg *short, straight*
10 a verb of motion, eg *walk*

2 (10 marks: 1 mark for each appropriate answer.)

1 drive
2 do
3 speak
4 buy/get
5 cook/make
6 see
7 ride
8 borrow
9 play
10 use

SECTION 2: GRAMMAR [30 marks]

3a (10 marks: 1 mark for each correct answer.)

1 c) they're
2 c) of
3 c) staying
4 b) opposite
5 c) was
6 b) go
7 a) visited
8 b) are
9 c) wonderful
10 a) by

b (10 marks: 1 mark for each appropriate answer.)
Possible answers

11 do
12 got
13 some
14 How/What
15 any
16 go
17 in
18 on
19 Have
20 out

4 (10 marks: 1 mark for each correct question.)
Possible answers

1 How do you go to school/work?
2 How long does it take?
3 How much is it?
4 How far is it?
5 How tall is he?
6 How old are you?
7 What's he like?
8 What does he look like?
9 What time do you get up?
10 Can you knit?

SECTION 3: READING [20 marks]

5 (10 marks: 2 marks for each correct answer.)

1 a shower
2 coffee
3 lunch
4 dinner
5 to sleep

6 (10 marks: 2 marks for each correct answer.)

1 DS 2 T 3 T 4 T 5 F

SECTION 4: WRITING [20 marks]

7 (20 marks)

Tell students what you will take into consideration when marking their written work. Criteria should include:

* efficient communication of meaning (7 marks)
* grammatical accuracy (7 marks)
* coherence in the ordering of the information or ideas (3 marks)
* layout, capitalisation and punctuation (3 marks)

It is probably better not to use a rigid marking system with the written part of the test. If, for example, you always deduct a mark for a grammatical mistake, you may find that you are over-penalising students who write a lot or who take risks. Deduct marks if students haven't written the minimum number of sentences stated in the test.

Answers Progress Test 3 Lessons 21 – 30

SECTION 1: VOCABULARY [30 marks]

1a (10 marks: 1 mark for each correct answer.)

1 eight
2 Monday
3 casual
4 coffee
5 black
6 bag
7 whose
8 medicine
9 fast
10 knitting

b (10 marks: 1 mark for each appropriate answer.)

1 an ordinal number, eg *second, fourth*
2 a month, eg *January, May*
3 an item of clothing, eg *jeans, T-shirt*
4 something to eat, eg *cheesecake, sandwich*
5 an adjective for describing an object, eg *heavy, large*
6 a material, eg *leather, metal*
7 a possessive pronoun, eg *hers, ours*
8 a complaint, eg *sore throat, stomach ache*
9 a superlative adjective, eg *dirtiest, oldest*
10 a sport, eg *boxing, tennis*

2 (10 marks: 1 mark for each appropriate answer.)

1 write/read
2 go
3 wear
4 do/get
5 spend
6 shake
7 buy
8 am/feel
9 drink/have
10 think

SECTION 2: GRAMMAR [30 marks]

3a (10 marks: 1 mark for each correct answer.)

1 b) I don't
2 c) some
3 c) you
4 c) you've
5 b) mine
6 a) did
7 c) on
8 a) a
9 b) of
10 a) because

b (10 marks: 1 mark for each appropriate answer.)
Possible answers

11 Can
12 for
13 over
14 on
15 ahead
16 me
17 not
18 biggest/only
19 in
20 it

4 (10 marks: 1 mark for each correct sentence.)

1 She hasn't got an aspirin.
2 He didn't find his coat.
3 They didn't have a happy marriage.
4 We don't like horseriding.
5 I'm not older than John.
6 He isn't wearing brown shoes.
7 You shouldn't sit down.
8 She doesn't do the housework.
9 It wasn't wet.
10 We aren't going away this year.

SECTION 3: READING [20 marks]

5 (10 marks: 2 marks for each correct answer.)

1 DS 2 F 3 T 4 DS 5 DS

6 (10 marks: 2 marks for each correct answer.)

1 (For the first time in 1968) a woman lit the Olympic flame.
2 197 (teams were at the Atlanta games).
3 (There are five rings) because there are five continents in the world.
4 The first modern Olympic games took place (about a hundred years ago).
5 (Beach volleyball first became an Olympic sport) in 1996.

SECTION 4: WRITING [20 marks]

7 (20 marks)

Tell students what you will take into consideration when marking their written work. Criteria should include:

* efficient communication of meaning (7 marks)
* grammatical accuracy (7 marks)
* coherence in the ordering of the information or ideas (3 marks)
* layout, capitalisation and punctuation (3 marks)

It is probably better not to use a rigid marking system with the written part of the test. If, for example, you always deduct a mark for a grammatical mistake, you may find that you are over-penalising students who write a lot or who take risks. Deduct marks if students haven't written the minimum number of sentences stated in the test.

Answers Progress Test 4 Lessons 31 – 40

SECTION 1: VOCABULARY [30 marks]

1a (10 marks: 1 mark for each correct answer.)

complaints	cough, headache
countries	China, Hungary
family	aunt, sister
food and drink	salad, water
furniture	chair, cupboard
jobs	journalist, waiter
means of transport	ferry, plane
months	December, March
shops	chemist, florist
sports	cycling, golf

b (10 marks: 1 mark for each appropriate answer.)

complaints	eg *cold, stomach ache*
countries	eg *France, Spain*
family	eg *husband, mother*
food and drink	eg *bread, pasta*
furniture	eg *bed, table*
jobs	eg *secretary, teacher*
means of transport	eg *bicycle, train*
months	eg *July, September*
shops	eg *newsagent, supermarket*
sports	eg *boxing, tennis*

2 (10 marks: 1 mark for each appropriate answer.)

1 take
2 stay
3 listen
4 pay
5 use
6 make
7 speak
8 pass
9 work
10 score

SECTION 2: GRAMMAR [30 marks]

3a (10 marks: 1 mark for each correct answer.)

1 b) I'll
2 a) How
3 a) at
4 b) so
5 a) ever
6 c) was
7 a) ago
8 c) went
9 a) are you staying
10 c) this

b (10 marks: 1 mark for each appropriate answer.)

Possible answers

11 were
12 just
13 was
14 Did
15 on
16 every
17 having
18 got
19 should
20 be

4 (10 marks: 1 mark for each correct expression.)

1 I
2 I'd
3 was built
4 had
5 takes
6 doesn't speak
7 was
8 haven't finished
9 hurt
10 I'm wearing

SECTION 3: READING [20 marks]

5 (10 marks: 2 marks for each correct answer.)

1 b 2 d 3 d 4 b 5 c

6 (10 marks: 2 marks for each correct answer.)

1 DS 2 T 3 T 4 F 5 F

SECTION 4: WRITING [20 marks]

7 (20 marks)

Tell students what you will take into consideration when marking their written work. Criteria should include:

* efficient communication of meaning (7 marks)
* grammatical accuracy (7 marks)
* coherence in the ordering of the information or ideas (3 marks)
* layout, capitalisation and punctuation (3 marks)

It is probably better not to use a rigid marking system with the written part of the test. If, for example, you always deduct a mark for a grammatical mistake, you may find that you are over-penalising students who write a lot or who take risks. Deduct marks if students haven't written the minimum number of sentences stated in the test.